PATCH ATLAS

Patch Atlas

INTEGRATING DESIGN PRACTICES

AND ECOLOGICAL KNOWLEDGE

FOR CITIES AS COMPLEX SYSTEMS

Victoria J. Marshall

Mary L. Cadenasso

Brian P. McGrath

Steward T. A. Pickett

Yale UNIVERSITY PRESS   NEW HAVEN AND LONDON

Frontispiece—Five land cover elements shown as fragments of aerial imagery: woody vegetation, herbaceous vegetation, bare soil, pavement, and buildings

Published with assistance from the foundation established in memory of Calvin Chapin of the Class of 1788, Yale College.

Yale University Press books may be purchased in quantity for educational, business, or promotional use. For information, please e-mail sales.press@yale.edu (U.S. office) or sales@yaleup.co.uk (U.K. office).

Designed by Nancy Ovedovitz and set in Scala Sans type by Tseng Information Systems, Inc. Printed in China.

ISBN 978-0-300-23993-5 (paper : alk. paper)
Library of Congress Control Number: 2019937271
A catalogue record for this book is available from the British Library.

This paper meets the requirements of ANSI/NISO Z39.48-1992 (Permanence of Paper).

10 9 8 7 6 5 4 3 2 1

CONTENTS

PREFACE: FOUR THEMES FOR AN ATLAS

An atlas is a system of related maps, images, and information about some particular place. This patch atlas is based on a new classification tool created to understand land cover in a region encompassing city, suburbs, and exurbs as a complex and hybrid social-ecological system. Presenting this new description of an urban region relies on maps of land cover classes and patches, tables showing the relationships of the various covers, information suggesting how the specific arrangements of cover came to be, and speculations on how those arrangements might change via design and other processes, such as disturbance and succession. Consequently, this atlas extends the new land cover classification tool into a system for design.

Although *Patch Atlas* does focus on a real watershed encompassing a swath of the Baltimore region, it is perhaps more important as a new conceptual approach appropriate for describing and envisioning change in any complex, urban place. The new conceptualization has four characteristics that underlie the logic that unfolds through the book.

First, the atlas is co-produced by urban designers and urban ecologists working together. All four of the authors have contributed through dialogs, urban design studios, scoping drawings, exploring new tabulations of urban cover elements, field trips to ground the maps across Baltimore, shared reading and discussion of precedents, and finally speculating together on the implications of the classification for design and urban change. Thus, there are no separate science chapters or distinct design chapters, and no summative application chapter. All of us—designers and scientists alike—have struggled with and contributed to the mapping, to the tabulations of pattern, to the graphical strategies for exposing relationships among covers, and to extracting meaning and implications of the new patterns we discovered. We thus hope to share a sense of what a new way to represent the integrated social and ecological structure of urban land cover can contribute to the scholarship of ecology and the practice of design.

Second, *Patch Atlas* has an unusual place in the development of urban land cover classifications. It contributes to a long engagement with the relationship of nature and culture that is explicit in the built environment. This is a huge area, and cannot be dealt with comprehensively in a short book. We use, however, key precedents in urban classification to help situate our joint work within the trajectory of improving urban classifications, but also to note others on the same path of classifying settled places as inherently both socially and biophysically generated. However, this atlas differs from most precedent land cover classifications, even those that have been widely recognized as promoting an ecological view of urban systems. In particular, *Patch Atlas* contrasts starkly with the predominant practice of starting urban classification by separating human-constructed covers from covers that emerge from predominantly biological and geological processes.

Third, *Patch Atlas* encourages speculative design. Although our maps, tables, and images represent a particular time in the recent history of Baltimore, such temporal anchoring immediately suggests questions of interest to both designers and those concerned with the ecological and social implications of design. How did the particular cover combinations come to be? What was the role of past design visions in creating the cover mosaic we map here? What do the different cover classes and their spatial configuration suggest about how desired urban transformations might be thwarted or can be promoted? In other words, *Patch Atlas*—in its maps, tabulations of cover patterns, and images relating the various cover types to each other—invites communities to speculate about "the city yet to come" and how their vision and work might better link culture and nature in the future.

Fourth, *Patch Atlas* can be relevant to disciplines and practices beyond those professed by the authors. It is also of interest to residents curious about the form and dynamics of their surroundings. Urban land covers are generated by many causes, and are the substrate upon which many professional practitioners act and in which residents live. Consequently, sociologists, geographers, anthropologists, and historians may have need of hybrid representations of urban places and their potential for change. Likewise, engineers, who are one of the main agents of built infrastructure in urban systems, work in the context of complex urban land covers such as those revealed by the representations in this atlas. Regional planners may find useful challenges in the classification. Biologists interested in the richness and kinds of plants, animals, and microbes that share the city with humans can also use the hybrid land cover classifications. Hydrology at the surface and below ground can relate to hybrid covers as well. And even those interested in environmental and social justice in urban systems may find useful indicators or potential drivers through the use of an atlas of this sort. All of these disciplines may be reflected in practices of restoration and revitalization that are so in demand in today's urban systems. While the book acknowledges connections with many other disciplines and professions, it necessarily emphasizes the disciplines its authors represent.

These four themes run throughout and help justify the motivations and choices we make. Of course, *Patch Atlas* and the new conceptual system it represents do not address the concerns of all urbanists. Some purposeful exclusions from our core conceptual structure include land use as a familiar classification goal; the actual activities that take place in particular areas; the third spatial dimension; social and demographic data; quantitative analysis of changes through time; and flexible or comprehensive building typologies. We hope that these and many other topics, which we cannot include here, stimulate future research and application of the conceptions and methods represented by the atlas to other urban areas and time horizons. We therefore invite readers to consider new research opportunities and application strategies.

ACKNOWLEDGMENTS

Baltimore Ecosystem Study Long-Term
Ecological Research, National Science
Foundation
Cary Institute of Ecosystem Studies
National Science Foundation Career Grant
Urban Sustainability Research Coordination
Network
Tishman Environment and Design Center,
The New School

Till Design, Victoria Marshall, Principal
urban-interface, Brian McGrath, Principal
Phanat Xanamane
Christian Neuner
Bryn Montgomery
Kirsten Schwarz

There are many reasons to be intrigued by cities, and to generate new ways to portray cities, and indeed any kind of urban area. We are motivated by the nature and importance of cities themselves, by their hybrid urban structure, their obvious and changing spatial patchiness, and by the opportunity we saw to develop and employ a novel land cover classification that could serve both urban science and urban design, as well as other scholarly and practical disciplines working in the urban realm. We hope the novelty of the way we model and depict urban land cover can benefit regional planning, engineering, social-economic sciences, architecture, ecological restoration, environmental justice, and other fields. This chapter lays out our motivations. It also introduces an analytical and theoretical approach to land cover that is embodied in the new classification, and presents a visual framework for understanding urban systems using the new classification.

Motivation for Understanding Cities

It has become commonplace to introduce discussions of urban systems with the recognition that such areas are now, and increasingly will be, the home of humankind. This fact and the trajectory it represents confirm the dedication of designers, such as architects and landscape architects, to better understand the urban realm. The urbanization of the planet also justifies the increasing attention of ecological scientists to urban areas, including cities, suburbs, exurbs, and the long-distance connections urban areas experience with other settled landscapes as well as wild and agricultural lands. Furthermore, urban areas—to use this term as a shorthand for all of the urban variety and complexity mentioned above—are exhibiting increasingly diverse and globally divergent forms and processes. No longer can simplistic labels such as city, metropolis, megalopolis, and the like be assumed to describe urban forms that are universal and locked into a single developmental sequence.

Cities are characteristically patchy across space, and there are many ways to understand such spatial heterogeneity. In fact, the pervasiveness of spatial heterogeneity within urban areas at various scales is a major motivation for pursuing new ways to examine and represent land cover mixtures. The heterogeneity of cities appears in their social ecological form, their extensive and patchy fabrics, their demographic structures and social connections, and their biological and ecological processes and outcomes. Spatial hybridity and heterogeneity are thus concerns shared not only by urban designers, architects, landscape architects, and ecological scientists, the home disciplines of the authors of this Patch Atlas, but also by engineers, planners, sociologists, political ecologists, and economists, among others.

Urban areas, in all of their variety, are a research frontier for the science of ecology. In general, the discipline of ecology seeks to understand the relationship between organisms and their environment. Not all ecologists avoid the city as a viable system to study, but the tradition in U.S.-based ecology has been to focus on places where the influence of humans appears to be minimal or is inconspicuous because it remains indirect. This shifted in the late 1990s when the National Science Foundation funded two long-term ecological research programs—one based in Baltimore, Maryland, and the other in Phoenix, Arizona—to study cities as hybrid social-ecological systems. The authors of this atlas are members of the Baltimore Ecosystem Study (BES), whose goal it is to understand the urban system of the Baltimore metropolitan area as an ecosystem.

The phrase, "understand the urban . . . as an ecosystem," embodies a key motivation behind the Patch Atlas. An urban ecosystem is necessarily a hybrid system, inextricably binding social processes and products with biophysical or ecological processes and products. This means, among other things, that cities will reflect the intentionality and creativity of design and planning, the metabolism and dynamics of organisms ranging from microbes to carnivores, and the purposeful management by agencies, organizations, and households. But some aspects and places in the city will also emerge from accidents, indirect effects of biological and social processes, natural disturbance, political catastrophe, the neglect of marginal sites, and invisible flows. As a result of this complex hybridity, the existence and care of urban areas invite cross disciplinary collaboration, research, and action.

Motivation for a Novel Urban Classification System

In spite of the complex hybridity of cities, urban heterogeneity is frequently described using simple land use classes. The classes reflect a siloed view of urban form where areas are catalogued as residential, commercial, industrial, or devoted to transportation, and so on. Depicting heterogeneity based on land use classes such as these embodies assumptions about human activities that occur in separated places and, based on those activities, deems areas internally uniform. Different areas classified as a given land use type may, however, differ greatly in urban form. Different areas of a given land use class may also combine the effects of social and biophysical processes in different ways. For example, areas identified as residential may differ from each other in terms of building size, type, and density, the abundance and type of vegetation, or the distribution and extent of paved and bare areas. All of these urban features—buildings, vegetation, and surface materials—may influence ecological processes differently. What are the links between the amount and distribution of various elements of urban form and ecological and social processes?

To address this fundamental question,

depictions of urban form based on land use are inadequate, and a different conception of urban heterogeneity is required. This limitation motivated us to develop a new tool to characterize and quantify the heterogeneity of urban form based on land cover. This new tool is called High Ecological Resolution Classification for Urban Landscapes and Environmental Systems (HERCULES), and it is described in detail below. Briefly, HERCULES distinguishes patches that differ from one another based on the mixture and relative abundance of five elements: buildings, woody vegetation, herbaceous vegetation, paved surfaces, and bare soil. These five elements of land cover can vary independently of each other in different situations. This means that adjacent patches may differ in the cover of any number of the five elements. Because HERCULES was developed for terrestrial systems, water is not included as an element, though it could be easily added if a specific research question required knowledge of the amount and location of water.

Addressing and using urban heterogeneity is a large and old pursuit. We briefly position HERCULES in this broad intellectual and practical space, recognizing that a comprehensive review is well beyond our scope. One of the seminal approaches to heterogeneity was that of Ian McHarg (1969), who pioneered the examination of heterogeneity within different "layers" of a landscape in order to identify optimal sites for new development. Layers address topography, hydrology, soils, vegetation, and historical factors, among others. Another approach to heterogeneity in urban areas is adapted from island biogeography theory and its expansion into terrestrial situations. First used by conser-

vation biologists, this view of a landscape recognized three kinds of sites: patches that act as habitat for a target organism; corridors that enhance dispersal among habitat patches; and a matrix of land inhospitable for the species, which therefore did not serve as habitat. This formulation deduces ecological process from the pattern of a landscape. In some cases that deduction may be correct, while in other cases it may be incorrect, so that the deduction demands empirical test. The patch/corridor/matrix formulation, as this view is called, is also limited by the assumption that flows of any interest occur just along the corridors.

HERCULES must also be positioned relative to two important architectural theories. Architecture has employed figure/ground and typomorphological classifications of urban form. These two kinds of architectural classification question the modernist notion that cities benefit from land use classification and the planned separation of functional uses. Both of the architectural methods carefully examine how historical transformations of urban form are fundamentally constituted of heterogeneous mixes of uses within urban blocks. Importantly, building uses can change over time without an external transformation of urban form. This critique of modern functionalism in architecture aligns with HERCULES in questioning the predominance of land use classification in planning. While figure/ground and typomorphological classifications are limited by their focus on classifying only the relationship between built and open space, HERCULES adds vegetation and soil to the urban mix.

Another thread in the understanding of spatial heterogeneity emerged from the growing availability of aerial and satellite images. With this growth, it became more common for researchers to describe more subtle contrasts in landscape structure. In an effort to standardize approaches, Anderson et al. (1976) established a widely used classification strategy that was intended to be applied nationally at coarse scales, on the order of kilometers, which matched the scale of the early imagery. This classification strategy was hierarchical, and at the top tier, or Level I, urban land is separated from "natural" land such as forest and agriculture. This binary approach ignores the reality that urban land is a mix of built and non-built elements. In addition, the Anderson classification is based on land use and often forms the basis of many other classification systems developed as resolution was gained in imagery. A finer scale approach to land use/land cover is the classification strategy provided by Ellis and colleagues (2000). This strategy is intended to be a hybrid linking both social and ecological aspects of a landscape. Each patch is characterized by both land use and land cover, thereby confounding the two. Each of these classification strategies has benefits and limitations, and therefore, before one is used it must be evaluated in light of the research question it will be used to answer.

HERCULES occupies a distinctive place in this universe of land classification schemes and the theories they imply. HERCULES focuses on cover, rather than land use. It does not a priori define patches by function, as does the patch/corridor/matrix formulation. It is a typomorphological description that includes built and natural landscapes as figures, grounds, and fields. It is a system that accepts that social and biophysical processes jointly generate differences in land cover. Therefore, it does not separate "urban" from "natural" covers. It is amenable to application at any spatial scale. These characteristics emerged from an empirical need.

A practical motivation for developing HERCULES as a hybrid classification scheme was the need to understand land cover as a part of the Baltimore Ecosystem Study. The traditional land use/land cover classification available for the metropolitan region was simply too coarse—both in terms of spatial resolution and particularly in terms of conceptual refinement—for a hybrid system. Rather than force our ecological questions into a coarsely conceived land use/land cover mosaic, we decided to start from scratch and develop a classification that better fit the theoretical goals of BES. Thus, HERCULES is based solely on land cover and delineates patches based on the relative abundance of the five elements using high-resolution spatial data. Such a cover classification can provide a fundamental basis for examining the effects of biophysical processes like the flow of water and nutrients in systems, the biological and social composition of different locations, and the changes in urban form and ecosystem processes across extensive urban areas.

The Patch Atlas uses greater Baltimore as the demonstration urban area for HERCULES. Baltimore is a complex urban region, exhibiting in some places unoccupied housing and vacant land, while other areas are growing and becoming more dense. In particular, the work of BES focuses on the spatial heterogeneity and ecosystem processes of the 17,150-hectare Gwynns Falls watershed. The watershed has several advantages as a focus of study. It spans Baltimore City and Baltimore County, and encompasses dense old residential neighborhoods originating in the nineteenth-century "walking city," as well as early twentieth-century streetcar suburbs, and contemporary car-based suburban and exurban development. It contains old industrial areas, and abandoned brownfields, which contrast with new business parks and big-box stores. There are large parks, tree-lined rights of way, domestic yards, a few remaining agricultural fields, and vacant lots. The social structures and processes in the watershed are diverse and reflect many intertwined historic factors, demographic shifts, and policy decisions. Finally, it is intensively sampled to measure stream water flow and quality, meteorology, soil chemistry, vegetation cover, and biodiversity, for example. Social data are also dense and temporally rich, including demographics, environmental attitudes and behaviors, and organizational networks, among others. The HERCULES classification helps to integrate these diverse data sources and dynamics, and to relate them to urban form. Recognizing historical contingency, multiple pathways, and the lack of defined end points for urban dynamics, HERCULES is an important tool for examining the many trajectories and relationships between urban form and processes.

Creating HERCULES: Characterizing the Hybridity of Urban Form

The goal of HERCULES is to capture the heterogeneity in urban form that is visible and identifiable from high-resolution aerial photos. As discussed in the motivations above, HERCULES is a classification system that categorizes the hybrid social-ecological

variation in land cover within an urban area. The output of HERCULES is simply a contiguous set of patch boundaries that outline areas that are similar to each other in internal composition. The boundaries of these patches are drawn as a polygon layer within GIS (Geographic Information System). HERCULES does not identify the abstract institutions and invisible social rules that contribute to shaping urban form. It does however, provide a spatial reference so that transformations of urban form over time could be identified by comparing patch layers from two or more time periods. For simplicity, this atlas only examines the cover for one period in time; the extensive work of a time series analysis is being conducted under other projects.

Using false-color infrared aerial imagery of sub-meter resolution, we drew the patch outlines as a digital layer on top of the images. False-color infrared images show features in colors other than they would appear to the human eye. This imagery is often used for interpretation because atmospheric haze does not interfere with the image, and some characteristics are easier to see. In these images, vegetation is red, and the shade of red gives an indication of density and vigor, such that lawns, for example, appear as a lighter red than the leaves of canopy trees. Buildings and pavement appear in shades of gray and black, or sometimes lighter colors depending on the specific materials, and bare soil is light green. The images used were all captured on the same October day in 1999. Although the atlas is based on imagery from a specific day, what is important for the generality of the approach is that this raw material was used to develop a broadly applicable

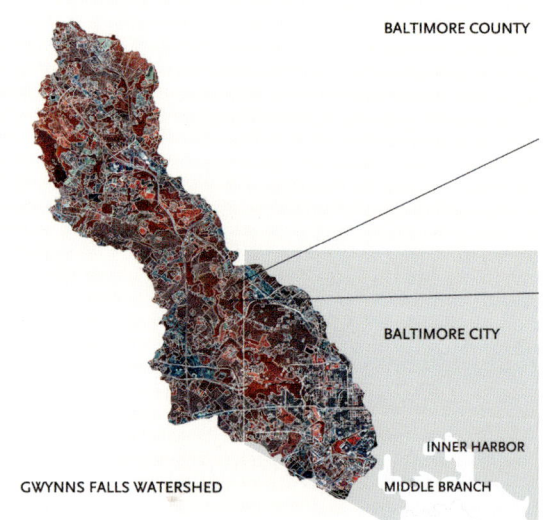

BALTIMORE COUNTY

BALTIMORE CITY

INNER HARBOR

GWYNNS FALLS WATERSHED   MIDDLE BRANCH

FIGURE 1 The patch array of the Gwynns
Falls watershed, at left, showing its location
in a map of the Chesapeake Bay, Baltimore
City, and Baltimore County. On the right,
a sample of one patch is shown in detail.

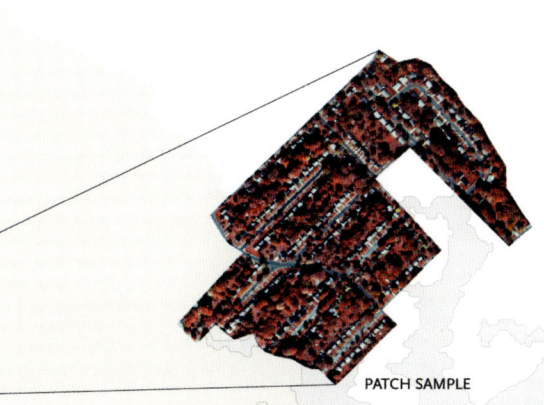

PATCH SAMPLE

2251 Total number of patches
in the Gwynns Falls watershed

342 Number of unique patch types
in the Gwynns Falls watershed

CHESAPEAKE BAY

0   3 miles   6

approach to urban land cover—one that has both theoretical implications and speculative design applications. Such applications must also be inclusive and participatory to ensure social equitability.

To identify the HERCULES patches, we capitalized on the exceptional skill of the human eye to recognize pattern and developed three primary rules to guide patch delineation decisions.

Rule 1. A new patch is drawn when the relative cover of any one of the five elements—woody vegetation, herbaceous vegetation, bare soil, pavement, and buildings—changes from one abundance class to another. Abundance classes are defined as absent (0), and present, at 10 percent (1), 11–35 percent (2), 36–75 percent (3), and greater than 75 percent (4).

Rule 2. A single patch must be at least 20 × 20 meters in two orthogonal directions. There were several specific reasons for this rule. First, 20 × 20 meters is larger than the average size of a residential parcel, preventing the design and management decisions made by an individual land owner from becoming a specific patch. Second, 20 × 20 meters is larger than a typical two-lane residential road, preventing each road from becoming its own patch, and instead incorporating the pavement of streets and local roads into the more complex patches they pass through. Areas with high road density would therefore be recognized as having greater pavement cover. Finally, some forests exist in Baltimore as remnant patches or parks, and these forests naturally have gaps in the canopy as trees fall due to senescence or storms. If that gap was smaller than 20 × 20 meters it was considered to be part of the surrounding forest stand, and woody vegetation cover was assigned for the entire patch. This is how ecologists usually treat treefall gaps in forests.

Rule 3. If a patch was bounded by a road, the boundary should be drawn through the middle of the road, allocating paved surfaces to patches on either side of the boundary. This rule was aimed at preventing biased allocation of this element and also is the reason that some patches that appear to be forests may have some pavement.

We found that different people were remarkably consistent in delineating patch boundaries. After patches were drawn, then the amount of each element inside the patch was estimated by the user and the patch assigned a string of 5 digits as the patch code. Each digit in the string represented the abundance class of the specific type of cover, as defined in rule 1. For example, if a patch was assigned the code 21022, the patch contained 11–35 percent cover of woody vegetation, 10 percent herbaceous vegetation, no bare soil, 11–35 percent pavement cover, and 11–35 percent building cover. Because the imagery is orthophotography captured perpendicularly from above, the cover of all elements in the patch necessarily summed to 100 percent.

Periodic Table of Potential Patch Types

Once we had been working with the HERCULES patch maps of Baltimore for some time, we realized that there might be important or useful patterns in the fundamental organization of patch types. Consequently, we created a table of a logically complete set of patches from the HERCULES classification, using the basic ideas of combinatorics. Combinatorics is a mathematical field that studies how distinct objects can be combined given some stated constraints. For this atlas, the distinct objects are the five elements of urban land cover, and the constraints are the proportions each is permitted to occupy in a defined area. This logically complete set describes all possible patch types that can exist in an urban area.

The logically complete set of HERCULES patches is akin to the periodic table of chemical elements that arranges elements based on the fundamental attributes of their atoms. The chemical elements show regular patterns based on the number of protons in their atomic nuclei, the configuration of electrons into discrete orbits around the nucleus, and their associated chemical properties. Here, it is only necessary to see the periodic table of chemical elements as a metaphorical guide for seeking pattern among the patch types of HERCULES. Further details of the periodic table of the elements are beyond the scope of this book.

Over the past seventeen years we have discussed and studied the resulting patch patterns in design and science studios, seminars, and conferences. In this atlas we interpret these patterns in four distinct chapters. This narrative is provisional, as HERCULES can be queried in other ways, other watersheds can be studied, and different times and places can be compared. Nonetheless, this atlas situates our understanding of heterogeneity in the Gwynns Falls watershed at that moment captured by the images.

The basic components of the atlas include: 1) swatches, also called patch types, to represent a combination of land cover elements that are present, 2) a periodic table of possible patch types, 3) suites of related patch types, 4) graphs of types and numbers of the actual patch types that exist in our Baltimore watershed, 5) maps showing the distribution of the suites of patch

types across the watershed, and 6) patch samples to illustrate key characteristics about each suite. There are also three aesthetic components: the flattened graph, the watershed boundary, and white space. The basic components are described below.

Periodic Table

The periodic table is a tabular arrangement that is logically complete and includes an ordering system that can be read horizontally and vertically. In the periodic table of HERCULES, land covers (also called elements) are located across the top of the table, in the following order from left to right: woody vegetation (dark red), herbaceous vegetation (light red), bare soil (light green), pavement (blue), and buildings (black/gray). These colors are derived from the false-color aerial imagery employed in the atlas. The patch types are also ordered from homogeneous (100 percent of one cover type) to heterogeneous (five types at 20 percent each), from top to bottom in the periodic table.

Swatches

The periodic table is made up of rectangles that are filled with the various colors representing the land cover elements. These rectangles are called swatches. Imagine a small sample of cloth that is intended to demonstrate the look of a larger piece of patterned fabric. Each swatch is one possible combination of elements and always adds up to 100 percent cover. The amount of the rectangle occupied by a specific color is the proportion of that cover in that patch type. Each swatch defines a specific patch type and is identified by a five-digit numerical code, described earlier. The order of the numbers in the code from left to right corre-

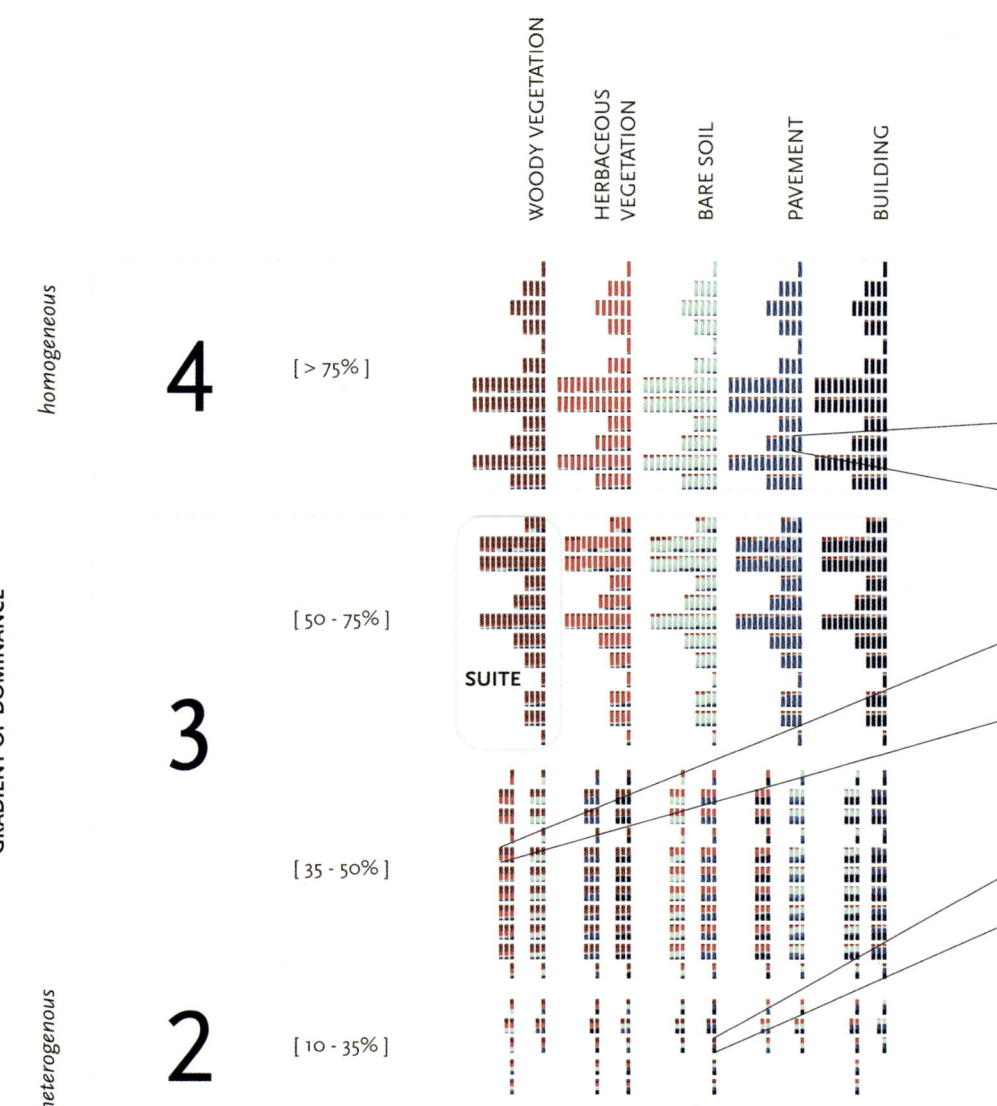

GRADIENT OF DOMINANCE

homogeneous

4   [> 75%]

3   [50 - 75%]

SUITE

[35 - 50%]

2   [10 - 35%]

heterogenous

FIGURE 2 The periodic table, at left, is composed of suites that are arranged as a gradient (vertically) and according to elements (horizontally). The diagram at right shows how the periodic table is also made up of swatches. Each swatch defines a specific patch type and has a nomenclature that is a string of five numbers. The numbers are read vertically in each swatch and they reference those numbers shown in the periodic table. For example, a number 4 in a swatch nomenclature means that one type of land cover dominates a patch, being present at greater than 75 percent. A number of 3 means that one or two types of land cover dominate a patch, with each co-present at 35–75 percent. A 0 means there is none of that land cover present in the patch, so it is not shown on the periodic table. It is impossible to have a patch that adds up to greater or lesser than 100 percent, so only the patches shown in the periodic table are possible.

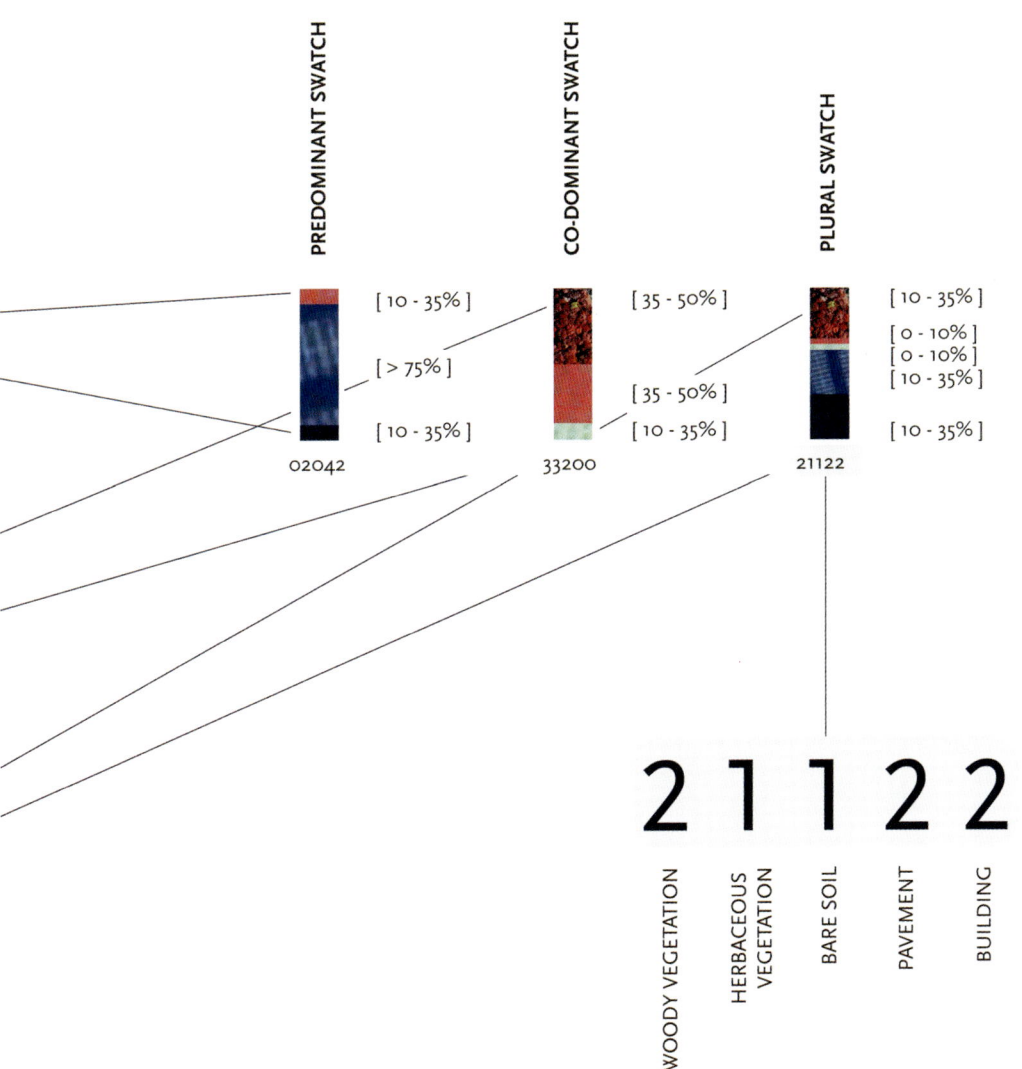

PREDOMINANT SWATCH

CO-DOMINANT SWATCH

PLURAL SWATCH

[10 - 35%]

[> 75%]

[10 - 35%]

02042

[35 - 50%]

[35 - 50%]

[10 - 35%]

33200

[10 - 35%]

[0 - 10%]
[0 - 10%]
[10 - 35%]

[10 - 35%]

21122

2 1 1 2 2

WOODY VEGETATION

HERBACEOUS VEGETATION

BARE SOIL

PAVEMENT

BUILDING

lates to the horizontal order of the periodic table: from woody vegetation to buildings. And the actual number correlates to the vertical order in the periodic table, which is from relatively homogeneous (4) to heterogeneous (1), and absent (0). In addition to representing the gradient from homogeneity to heterogeneity, the numbers 4 through 0 also identify breaks in the logic of combinations, so that 4 means greater than 75 percent, 3 means 36–75 percent, 2 means 11–35 percent, 1 is 10 percent or less, and 0 means the feature is absent.

Suites

The swatches are organized into suites. Suites are formed by breaks in the table that are changes in the dominance of the combination of elements. The different suites are discussed in Chapters 2 through 5.

Graphs

Two types of graphs are used to depict the number of each patch type actually found in the Gwynns Falls watershed. These graphs are an exact overlay of the periodic table and differ from each other in dimensionality—one is a three-dimensional depiction and the other is a flattened, two-dimensional version. In general, more clusters of color mean more patches are present, and the longer bars of color are a dimension that correlates to the actual number of patches. More specifically, the colors of the vertical striping reveal the swatches that are present, and the length of the bar reveals the numbers of patches in the Gwynns Falls watershed that are of that patch type—the patch frequency.

Map

For each patch type suite, a map is constructed by extracting the instances of

that type from a map of all patches in the Gwynns Falls watershed. This complete map shows every patch boundary and patch actually present in the Gwynns Falls watershed in October 1999. The boundaries can be seen as white lines. This complete map has all patches "turned on." As you page through the atlas, you will see that the subsequent maps only show patches that belong to a particular suite, leaving the rest of the watershed as white space. The maps reveal that the suites have different spatial patterns. The complete map and the maps for each suite type are GIS-based images, which means they are data-based drawings and the spatial location of each patch in the watershed is geo-referenced. Rather than a layer-based drawing that is created using illustration software, patches are turned on and off by querying the periodic table, displaying the result, and then saving them as an image.

Patch Sample

In each chapter, you will find a zoomed-in sample patch from the map of each suite. For both the map and the patch sample, there is a lot of white space, so they seem to float on the page. We did not turn on adjacent patches, adjacent watersheds, or the watershed boundary. This is a visual device to focus attention on something that is actually quite complex: the relationship of patches within and across the suites and individual patch types. The suite map is shown with all of its intricate details, and the zoomed-in sample allows you to look carefully and deeply at one patch at a time. In this way, white space is used as information. Just like the out-of-frame actors in a movie, it creates the circumstance, and focuses our attention on the situated action of the scene.

Motivation to Make an Atlas

This is not your typical atlas. Our motivation to create this Patch Atlas is to share our research as it stands now. It is not intended to pre-codify the world and to enclose space. Rather it aims to inspire maps that have yet to be drawn—to provide a set of rules for many atlases of the future. It offers a way to fill in the map of any urban watershed or region as a mosaic of integrative hybrid patches. Patch maps are a critical tool that can be used to speculate about the changes in cover that may result from specific designs, or the interaction of several proposed designs. Such speculative use is important in allowing design professionals, interdisciplinary teams, policy makers, and communities to communicate with one another in comparing and evaluating scenarios of change in urban areas.

In addition, the atlas works on another register. The spatial understandings of Baltimore have been inscribed in our individual consciousness as personal atlases of sorts. We extract, and reflect on the experience and insights we have had together in Baltimore. Hence, the atlas is a repository for more than seventeen years of shared knowledge about Baltimore. Through fieldwork, drawing, teaching, writing, reading, and talking we have amassed much knowledge about the people and neighborhoods of the city.

Finally, we know that HERCULES is complex, and that it takes time to explain before it might become useful. For this reason, we have framed the atlas as a journey. It is designed to orient the reader step by step into a view that feels disorienting in a productive way. The journey is not directed by a map, but rather by the periodic table. We

will move suite by suite from the relatively homogeneous top of the table down to the last suite of maximum heterogeneity.

How Does the Atlas Advance Design and How Does It Advance Ecology?

The atlas advances urban design in that it integrates both architectural and landscape approaches to the city. Although integration within design is a positive step, it can benefit from an ecological perspective. The most common use of ecology in design is as a metaphor. Used as a metaphor, ecology and its key components can stand for a variety of sometimes contradictory assumptions about patterns and processes in urban areas. Contrasting and contested terms such as crisis, emergence, connectivity, boundedness, stability, dynamism, metabolism, adaptability, resilience, and resistance evoke supposedly ecological ideas about cities used in design. These and other terms are used to label different kinds of urbanism—that is, ideologies and agendas for institutions, urbanists, landscape architects, architects, and urban designers. While there is no denying the power and utility of metaphor, the atlas is a tool to delve into substantive ecological knowledge about urban systems. Although the atlas uses Baltimore as its focus, HERCULES, its periodic table, and the actual roster of patch types and configurations can all be used to improve critical design discourse in other cities and urban areas.

Ecological science is an evidence-based approach to understanding "the processes influencing the distribution and abundance of organisms, the interactions among organisms, and the interactions between organisms and the transformation and flux of energy and matter." The key aspects of this

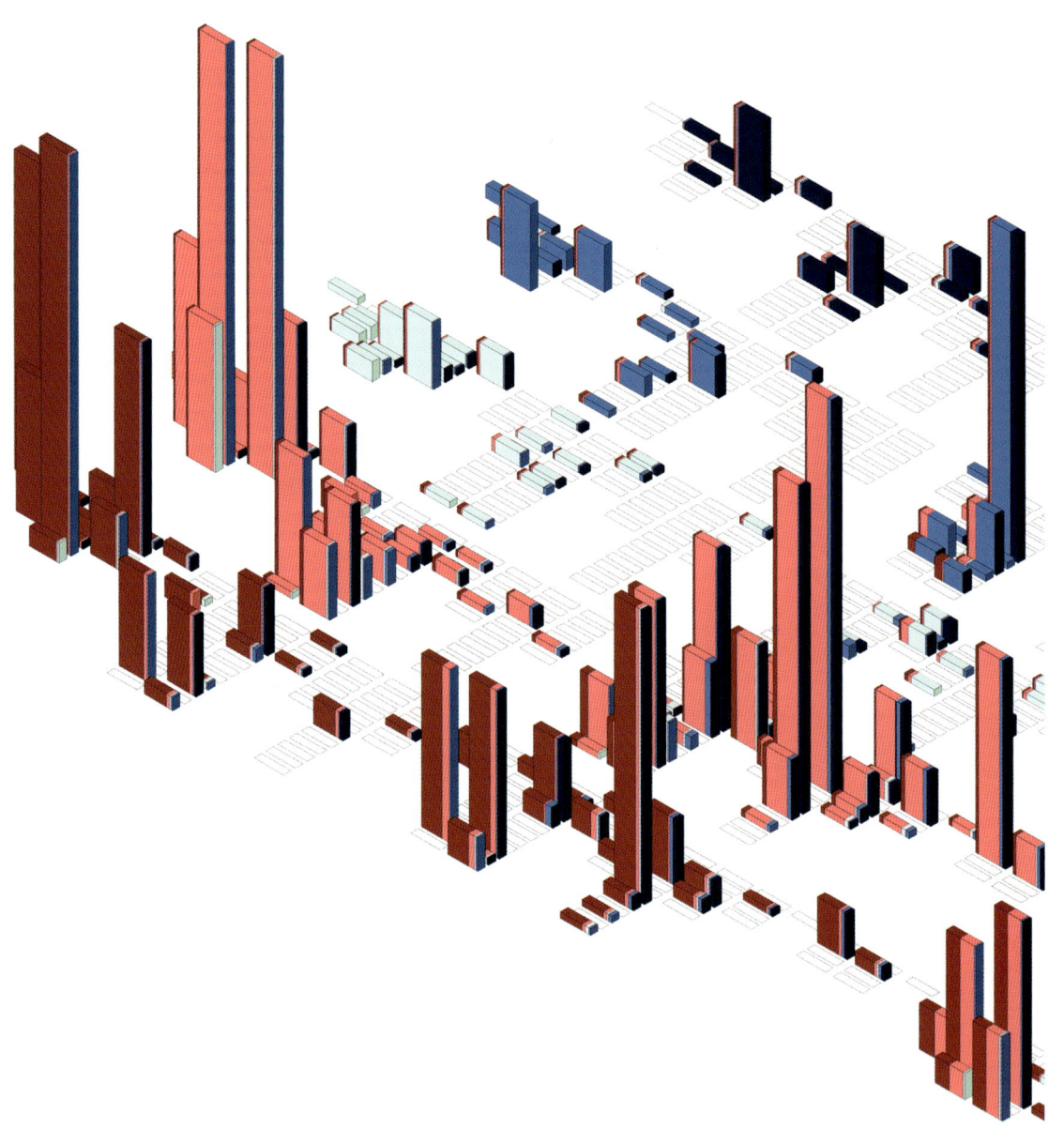

FIGURE 3 Three-dimensional graph or
"signature" of the patch array of the Gwynns
Falls watershed in Baltimore. Each patch swatch
increases in height in correspondence to the
number of those patches in the watershed.

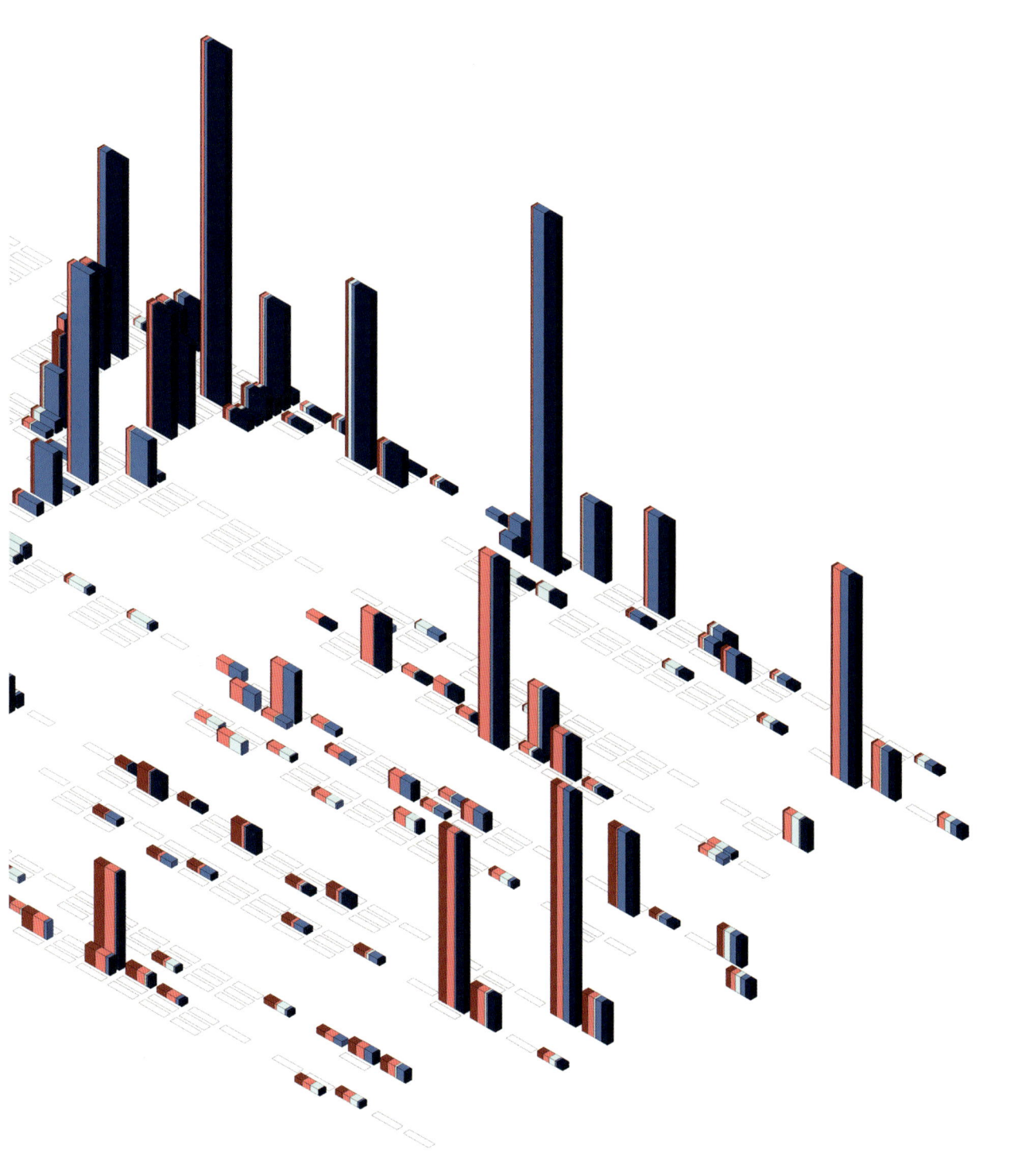

definition are a focus on the role of organisms and their interactions in structuring systems and driving the metabolic transformations of matter and energy. An ecological approach to urban systems thus focuses on what non-human organisms are present; how they function; how they interact with one another, with people, and with the physical environment; what pathways exist by which organisms effect the transformation and movement of nutrients, pollutants, contaminants, disease, and genetic information. Over time, the ecological approach to urban systems has matured into a joint ecological-social, interdisciplinary science that takes the human, social, constructed, and technological aspects of the system to be inextricably entangled with the biological processes and potential in cities, suburbs, and exurbs (CSE). Designers and other individuals make choices that influence all these social-ecological structures and interactions. Thus, design always intentionally or accidentally makes or modifies ecosystems, both at the scale of individual practice and at larger scales in which specific projects and interventions are embedded. Landscape architects, architects, urbanists, and planners engage people, affecting ecological-social structures and processes, and their social-ecological connections to larger or distant systems. Understanding this role may start with metaphor, but it must engage data, confirmed generalizations of observations, and models of the ecological structure and functioning of urban areas.

Ecological knowledge extends the interdisciplinary concerns of designers well beyond their usual scope informed by economic value, regulatory context, social program, aesthetics, and innovation. Ecology does not replace these facets of design. Rather it complements them. The spatial heterogeneity exposed by the Patch Atlas is a shared foundation for dialog between ecological science and urban design. Because the atlas starts with the assumption that urban heterogeneity is co-produced by social and biophysical processes, the maps and representations comprise patterns that can engage design and science practices as inter-referenced. The atlas allows designers to speculate about the effects of their proposals on the joint social-ecological form of an urban area.

The atlas helps deepen the role of ecology relative to design. Ecology is often part of the background of the development process. Ecological and environmental information are brought in toward the beginning of the design process as contributions to site analysis, and sometimes as part of visioning exercises with clients and stakeholders. The atlas reminds us that physical designs that are implemented have the capacity to alter patch structure through each of the three features that the HERCULES classification considers: vegetation, buildings, and surfaces. Thus, an important design application of the atlas is its use in situating particular designs and projects in an integrated context represented jointly by the three kinds of elements HERCULES employs. In addition, design is engaged as a participatory and inclusive process with stakeholders, in order to change policy and alter undesirable projects. Therefore, constructing HERCULES-derived patch maps that project the kinds of changes likely to appear after an intervention or initiative can help identify environmental changes to follow, and to share those projections with both the powerful and the marginalized. Such uses are a speculative, and normative, application of the atlas.

Patch maps can encourage joint evaluation of the successes or failures of a completed design project at meeting its goals over time. The contextual contributions and outcomes of a project can be identified using high-resolution patch maps, both before and after designed action. There are many ways to keep ecologists and communities involved in design practice, but updating and discussing the changes that can be captured in an evolving patch map are one stimulus to continue the dialog. Key here is recognizing that HERCULES patch maps can be constructed for different times. In addition, examining change in the mosaics over time can be used as descriptive, speculative, and inclusive tools.

Ecology in its turn is advanced by the atlas. The "periodic table" of potential patch suites and the frequencies of actual patches in a study area both act as fodder for research. Indeed, the periodic table of HERCULES patches can be seen to be a theoretical device of a sort rarely identified in land use/land cover research. The periodic table of land cover identifies two dimensions of differentiation—richness of cover elements, and evenness of how they are combined. Such contrasting conceptual dimensions are found in ecological theories of plant and animal community dynamics, or in landscape heterogeneity. New research can be generated following up on many of the trends and associations noted in the atlas. For example, the patterns illustrated in the atlas focus on patch type and number of each type. Chapters 2 through 5 will detail these patterns for different suites of patch types. The next generation of research can begin by addressing the details of patch patterns, including individual and aggregate area occupied by various patch types. Patch shape and dimensional complexity are also of interest, as are distances among patches of a type or among patches representing very different suites. How patch type, characteristics, and configurations have changed through time, or how they may change into the future, are compelling research questions, as suggested by the discussions in several chapters about design, demographic, and economic shifts that have led to the mutable patterns presented here. However, the goal of this atlas is to set those questions up by exploring the structure and implications of HERCULES patch types as general theory and as specific descriptions of a part of the Baltimore urban region at a particular time. Fleshing out the research agenda that emerges from that consideration is the task for future work and publications.

HERCULES has also been presented since its inception as a stage for examining drivers, outcomes, and controls on other kinds of structures and on interactions. If urban change is the "play," then HERCULES describes the physical setting in which that play unfolds, and provides a mechanism for elaborating the changes that the play generates with the setting. By incorporating cover elements that originate from both social and biological processes, HERCULES contextualizes the play in a way different from standard land use/land cover classifications. For example, the distributions and abundances—both number and area—of patch types and suites can be investigated relative to the details of such physical features as topography or to the presence, absence, or burial of streams and pre-urban drainage networks. Social characteristics, such as the classical ones associated with education, income, density, and ethnicity as well as more nuanced social features such as life-style groupings, social institutions, and governance networks can be intersected on the patch arrays presented here to generate new interdisciplinary hypotheses about urban areas as hybrid and heterogeneous social-ecological systems.

Finally, ecology can be advanced as a contribution to design practice by employing the Patch Atlas as a communication tool with other disciplines and various constituencies. Such uses emerge from the power of maps as a medium that most people are familiar and comfortable with. Even lay people are often fascinated by maps, and by aerial images of places they know. Unfurling a map, figuratively or literally, is an effective invitation for conversation and dialog. We have focused on the dialog between ecology and design, because those are the perspectives the authors represent. However, the power of maps as "boundary objects" with many other disciplines and stakeholders is enormous.

What Is New About It?

Perhaps the most novel feature of this atlas is the fact that it was co-created by designers and ecological scientists in a context of long-term ecological research investigating cities as ecosystems. The project represents an attempt to generate something that both specialties would understand and value. Neither discipline is viewed as being in service to the other, because all authors struggled with and contributed to the mapping, quantification of pattern, interpretation of new patterns, and strategies for

graphically presenting relationships among covers. As a result the social and ecological processes that shape built aspects of architectural and engineered components, and vegetation ranging from wild to manicured, all are included as structuring elements of an urban area. The design of the Patch Atlas reflects this integrative thinking. It is distinctly different from maps of heterogeneity that emphasize the built, on one hand, or the green on the other.

This integrative thinking helps both ecologists and designers to see an urban area, and space in general, as a co-produced hybrid of built and biophysical phenomena. Co-production in this usage refers to the urban system itself and not only to the more familiar idea of research agendas being jointly produced by people from different academic disciplines, professional specialties, or social sectors in a city. Co-production is based on the understanding that human-directed ecological change today cannot happen without concomitant social change. The heterogeneities illustrated in the atlas are true hybrid structures, co-produced by social processes and both managed and unmanaged biophysical structures and transformations.

The atlas can help translate this novel perspective of the hybrid social-ecological city from its scientific/design origins to other communities. Maps and atlases have a long tradition as public tools and communication devices. It is well known that colonial-era maps of the world, such as the Mercator projection, were used to visually privilege the colonial powers. Similarly, classical conservation maps show such preserves as national parks as distinct and separate from human society. And the redlining maps of the federally chartered Home

Owners' Loan Corporation from the 1930s were used to exclude racial minorities and immigrant populations from the benefits of insured mortgage lending. In contrast, this atlas does not reflect a colonial power agenda, or a restrictive division of nature from society, or an exclusionary racial segregation. Rather, it seeks to provide a critical and open-ended platform for comparison, discussion, and assessment of social and ecological goals.

The atlas helps people engage with their urban environments. Too often cities are seen as opposed to nature, or as having natural processes restricted to only large preserved green or blue spaces. Similarly, human societies are often seen as subjects that can be studied independently of their environment, and humans are often understood as entities whose subjectivity is found inside each person rather than in relation to an outside. The periodic table of hybrid patch types helps people—residents, experts, designers, policy makers, and others—see and discuss their city-suburban-exurban areas through a lens that blurs social, biophysical, and built together.

The boundaries between patches in the atlas are intended to reflect lived recognition of heterogeneity. In fact, we developed HERCULES, on which the atlas is based, as an antidote to the fact that we found the standard urban land classifications to be so devoid of social import and ecological nuance that we felt compelled to start from scratch. We were convinced that we were on the right track when our colleagues found the maps to reflect the ecologically relevant structures they were looking for as they organized their field sampling plans. Our intent was to capture the hybridity and heterogeneity that we saw on the ground, or

that emerged in discussions with residents, communities, and institutions in Baltimore. This would provide a tool to help ecologists and social scientists understand the pattern/process interactions in the watershed, but also communicate otherwise invisible or inconspicuous ecological functioning to residents, and others.

How to Read This Atlas

This chapter has laid out the motivations for the atlas from theoretical and practical perspectives and reinforced the four themes explored in the preface: 1) the atlas is co-produced by urban designers and urban ecologists, with neither discipline in service to the other; 2) the atlas applies a classification strategy that integrates socially and biophysically generated patterns; 3) the atlas is intended to encourage speculative and inclusive design; and 4) the atlas, though generated by urban design and urban ecology scholars, can be relevant to many other disciplines and stakeholders. The Patch Atlas describes the HERCULES system by which the heterogeneity of our focal region is examined. It has set out the flow that each subsequent chapter will follow, and has explained how the components of the Patch Atlas relate to each other in the remainder of the book.

Chapter 2 begins the detailed examination of the patchiness of the urban system by examining patch types that are dominated by only one of the five land cover elements. Thus, it focuses on the relatively more homogeneous land cover types in the watershed, and reveals a surprising role of vegetation in defining homogeneity in certain parts of the Baltimore region.

Chapter 3 opens the door to more internally heterogeneous patch types. Although

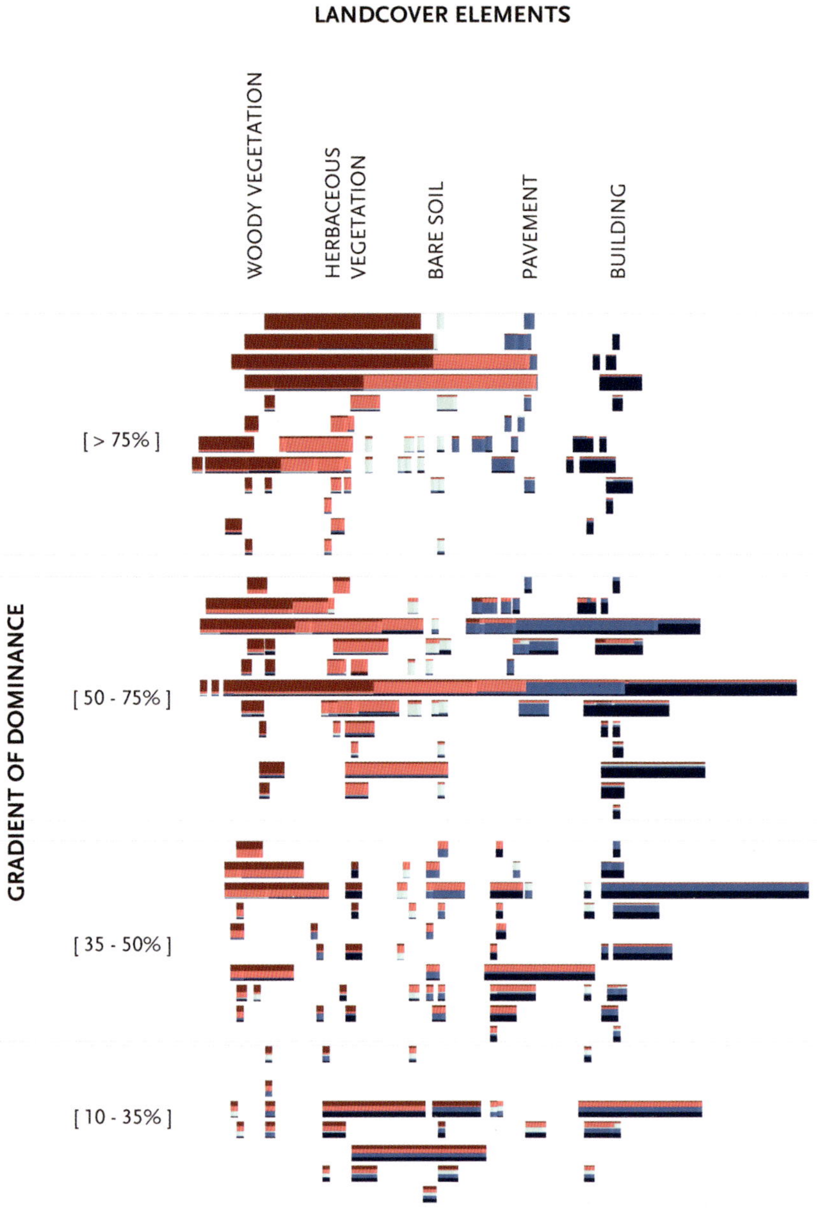

LANDCOVER ELEMENTS

WOODY VEGETATION

HERBACEOUS VEGETATION

BARE SOIL

PAVEMENT

BUILDING

GRADIENT OF DOMINANCE

[> 75%]

[50 - 75%]

[35 - 50%]

[10 - 35%]

Chapter 1
Motivations for characterizing
the hybrid, social-ecological city

Chapter 2
Unravelling homogeneity:
One predominant land cover element
with constrained potential for mixture

Chapter 3
Heterogeneity as outcome
of urban transformation:
One predominant land cover element
with greater potential for mixture

Chapter 4
Regularity within patches
as a characteristic of heterogeneity:
Two co-dominant land cover
elements and repeated pairs

Chapter 5
The case of patch plurality
as a lesson for urban mutability: Three
to five co-dominant land cover elements
and potential for recombination

Chapter 6
Speculating on urban futures

FIGURE 4 A flattened, two-dimensional version of the graph of the patch array of the Gwynns Falls watershed in Baltimore, left, with the table of contents for the Patch Atlas at right. The chapters are aligned with the graph to illustrate how the atlas is designed as a journey through the graph, from top to bottom, from homogeneous to heterogeneous, and across the top from woody vegetation to buildings.

one cover type remains dominant in this part of the periodic table, establishing a lower limit for dominance makes greater mixture of land cover types possible. A broader range of patch types actually appearing on the ground represents more of these kinds of patch types than the more strongly dominated types of Chapter 2.

While the earlier chapters unravel the concept of homogeneity and the way that urbanism shapes heterogeneity, Chapter 4 explains co-dominance of cover elements in detail. This chapter shows that certain pairs of cover elements often appear together in the Baltimore region. These pairings appear to relate to parcelization and commercial strip development, or to various stages of the construction and maturation of suburbs.

Chapter 5 focuses on plurality of land cover elements in defining patch types. This condition also invites an inquiry on mutability. These extremely heterogeneous patches are rare in the watershed, but they may be on the threshold of shifting predominance of cover elements. Therefore, this chapter focuses attention on the cumulative small changes by a range of urban actors and how they may collectively make big differences.

The journey guided by this Patch Atlas leads ultimately to a critical and open-ended view of urban areas as a heterogeneous phenomenon, and a shifting mosaic of cover types, in which distinct elements are modified by design or happenstance, and which respond to legacy and intention. In other words, heterogeneity is both an outcome and a driver of land cover change. The atlas is also a view of inclusive, ecological urban design as an equal partnership among disciplines, and with communities. The concluding chapter frames ten thoughts for ecological urban design as an approach to advance the hybrid urban realm.

CHAPTER 2   UNRAVELLING HOMOGENEITY : ONE PREDOMINANT LAND COVER ELEMENT WITH CONSTRAINED POTENTIAL FOR MIXTURE

Familiar land use classifications can delineate urban areas into homogeneous patches labeled as residential, commercial, and industrial, for example. Those patches are each identified for a particular use; however, within a single use class, the patches may differ tremendously in terms of actual cover. For example, some residential neighborhoods may have tree-lined streets and yards while others may be densely built and lack trees. Despite this variation in the physical cover of the land, all of these areas would be considered residential. Viewing all of those patches as homogeneously residential misses the opportunity to recognize and work with the heterogeneity that is present in the land cover and flattens the ecological complexity of the system.

In this atlas, we focus on land cover rather than land use. This focus brings forward the heterogeneity that characterizes urban systems and recognizes the potential importance of the abundance and arrangement of urban land cover elements for social and ecological processes. Despite the spatial complexity of land cover that charac-

terizes urban areas, we start our exploration of the Gwynns Falls watershed by first focusing on areas of relative homogeneity. In the periodic table of all possible combinations of land cover elements, there are five suites of patches that contain more than 75 percent of a single land cover element. Though the patches primarily comprise a single land cover element, another land cover element or combination of elements can constitute up to 24 percent of the patch.

Homogeneous patches are relatively unusual on the ground. The relatively homogeneous patches that actually appear in the Gwynns Falls watershed are most frequently dominated by vegetation, both woody and herbaceous, likely because Baltimore is located in the deciduous forest biome and vegetation tends to be associated with constructed areas as in many North American cities. In contrast, fewer patches are dominated by pavement or buildings, a reflection of the cultural preference for tree-lined streets in American cities. A predominance of bare soil in a patch is relatively rare, and this patch type likely characterizes areas in the process of transforming from a dominance by vegetation to a mix including buildings and pavement. At any one time, patches in this transitory state are expected to be few.

FIGURE 5 An example of land in the Gwynns Falls watershed that is predominantly covered with woody vegetation.

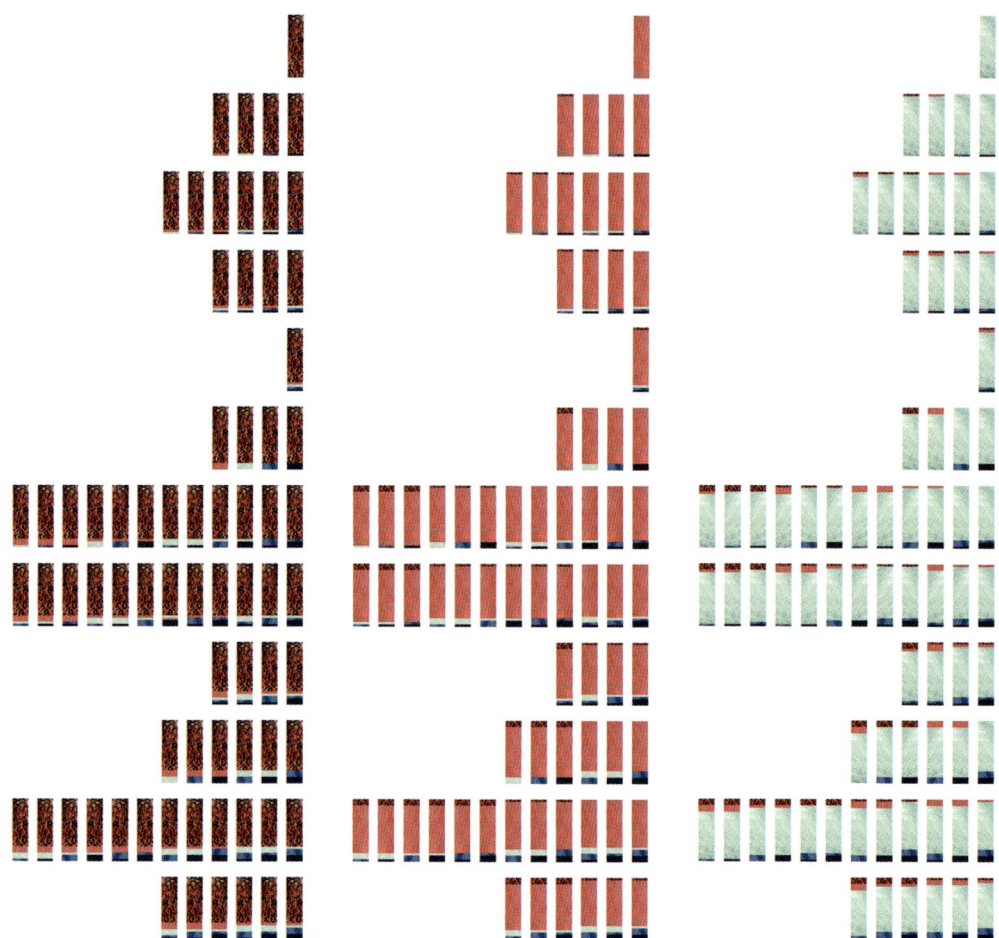

During the eighteenth and nineteenth centuries, forests declined as trees were harvested for timber and fuel and the land cleared for farming throughout the Gwynns Falls watershed. Therefore, patches dominated by trees today consist of regrowth and are associated with parks or conservation areas, areas unsuitable for building because of steep slopes and unstable soils, sites susceptible to flood risk, or parcels previously developed but where buildings have been removed and management discontinued. Patches predominated by herbaceous vegetation may include recreation fields, institutional grounds, golf courses, cemeteries, agricultural lands, and vacant lots. These

patches require management to maintain them in an herbaceous state. If unmanaged, they may fill with volunteer trees and transition into the patch type predominated by the woody vegetation that can be supported by the prevailing climate.

Woody Vegetation
The distribution of the specific patches of this type actually found in the watershed suggests that cover by a forest canopy is extremely frequent with only very small amounts of other land cover elements present. These woody vegetation patches may contain herbaceous vegetation in gaps among the trees that are smaller than the

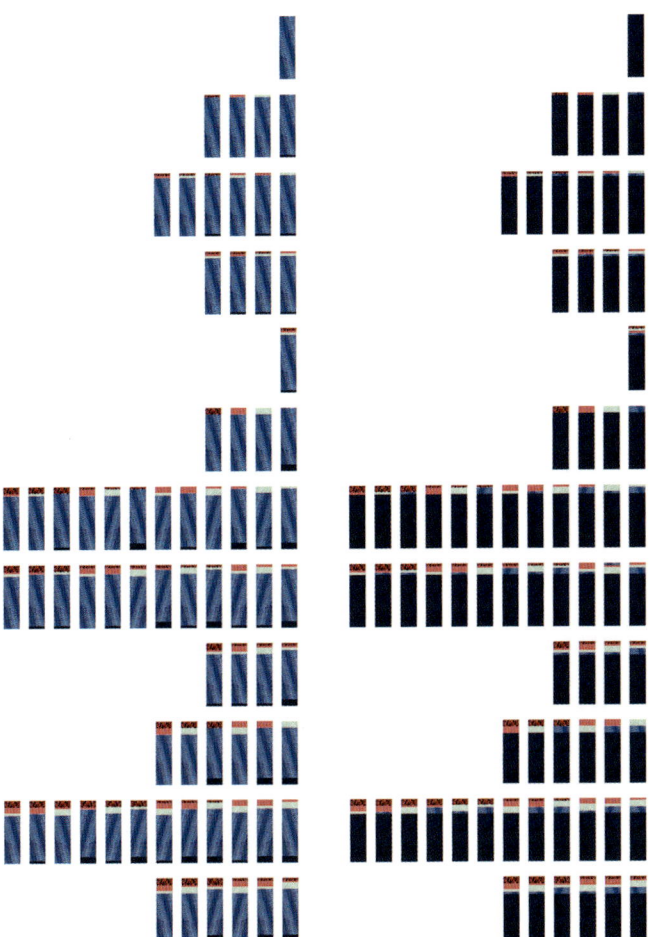

minimum mapping size of 20 × 20 meters, low-density residential areas, or a paved road that traverses the patch, for example. Woody patches are distributed throughout the watershed; however, in the lower watershed they are clustered along the stream valleys, while in the upper watershed the patches tend to be smaller and more dispersed. This contrasting distribution of large and smaller woody patches reflects a distribution that undoubtedly shaped the Olmsted Brothers' master plan for parks along stream valleys in Baltimore. The legacy of that public gesture is evident in the lower half of the watershed. In contrast, in the upper watershed the large woody patch

FIGURE 6 Overview: Periodic table of possible patch types with greater than 75 percent dominance by one type. The land cover elements from left to right are woody vegetation (dark red), herbaceous vegetation (light red), bare soil (green), pavement (light blue), and buildings (dark blue). The patch types at the top of the periodic table follow a gradient and are relatively more homogeneous than those lower down the figure. The five patch types that form the row at the top of the page are homogeneous, and these are very rare on the ground. The remainder of the patch types are composed of some mix of land covers, with a predominance greater than 75 percent of one type of land cover.

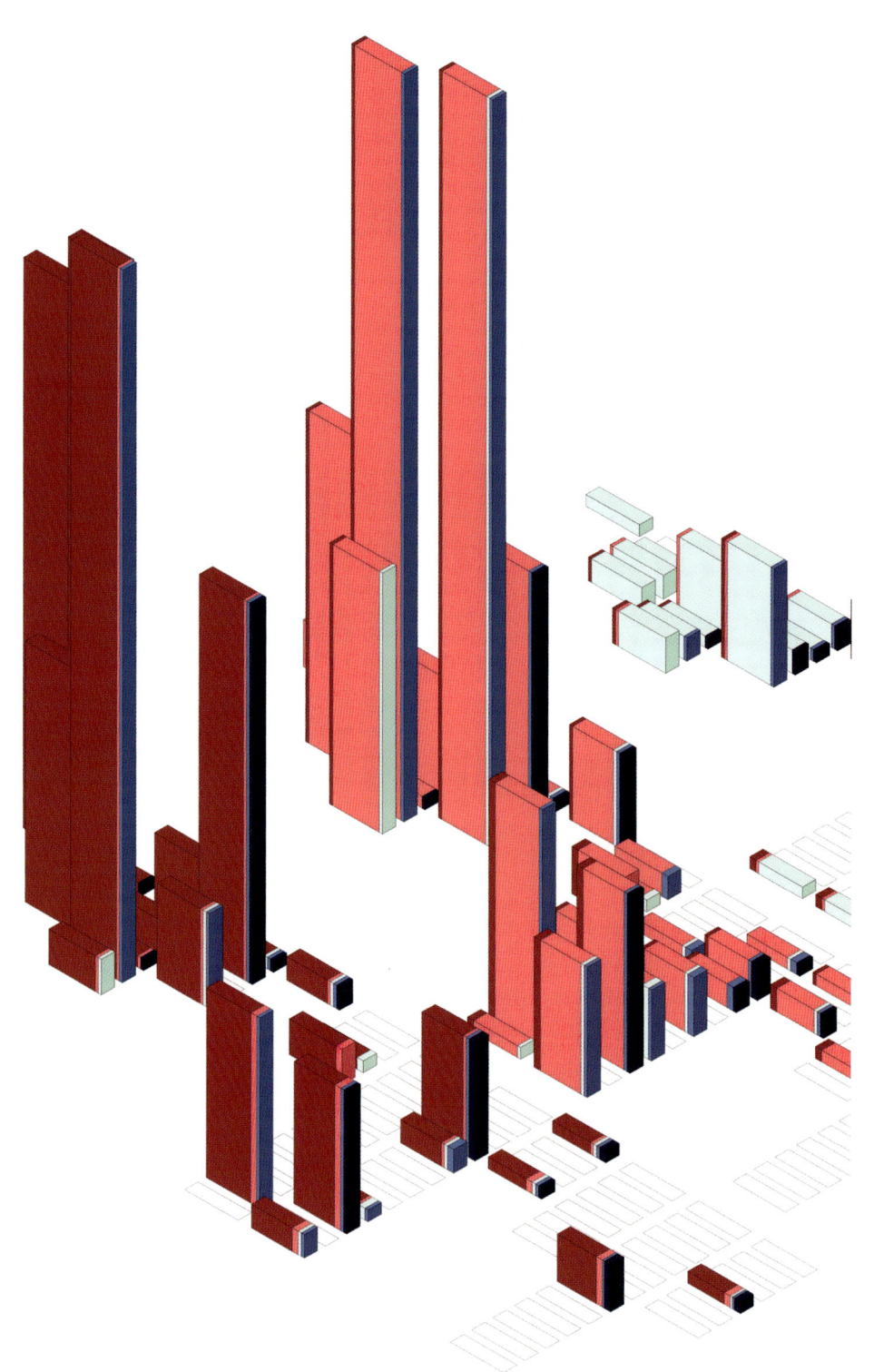

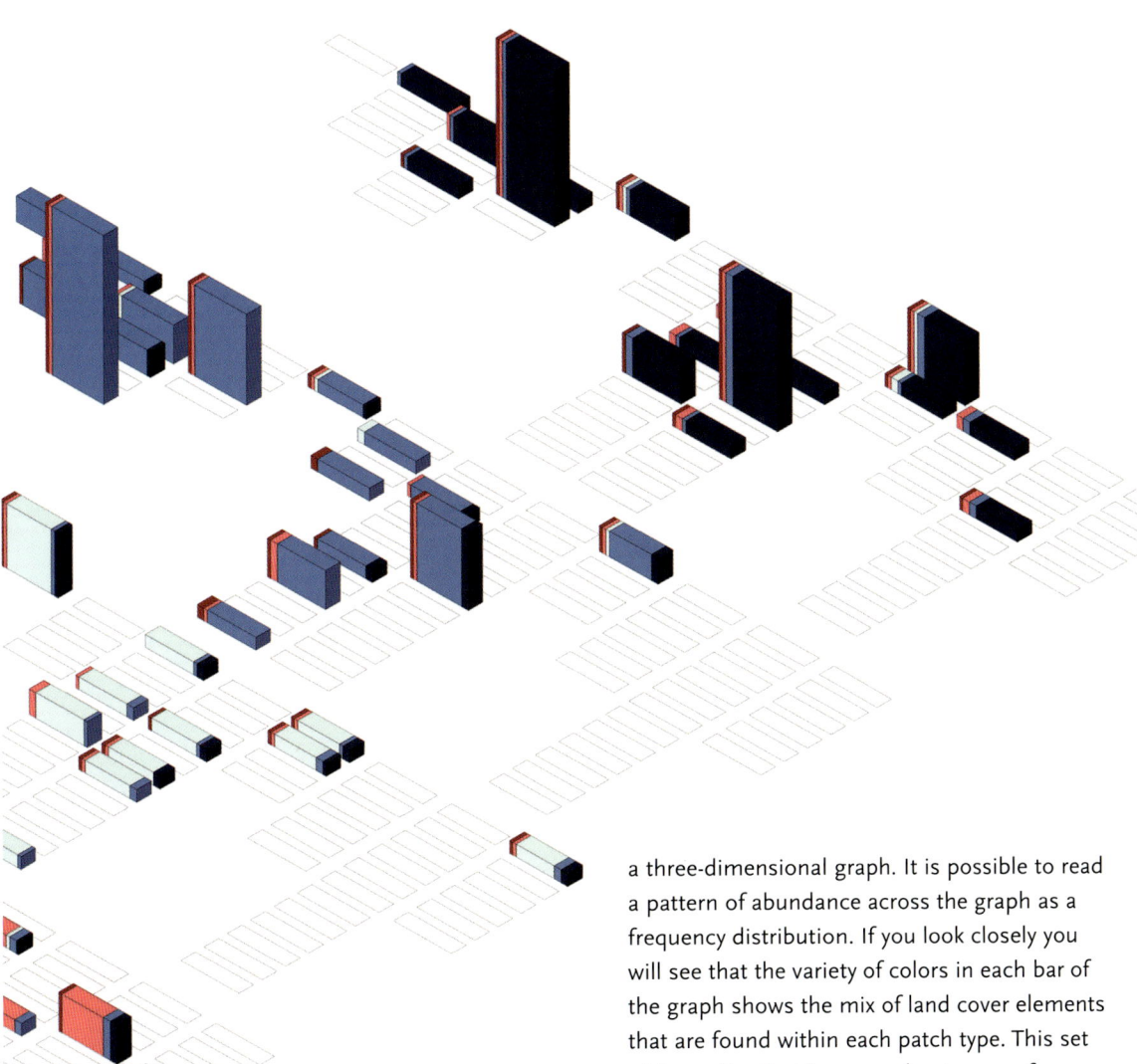

FIGURE 7 Overview: Graph of actual patch types in the Gwynns Falls watershed with greater than 75 percent dominance by one type. This three-dimensional figure is based on the periodic table of possible patch types (gray boxes); however, it only shows the patch types that are actually present in the Gwynns Falls watershed. The number of patch types that are present are shown vertically, forming a three-dimensional graph. It is possible to read a pattern of abundance across the graph as a frequency distribution. If you look closely you will see that the variety of colors in each bar of the graph shows the mix of land cover elements that are found within each patch type. This set of five suites that has a predominance of one type of land cover in a category of greater than 75 percent shows a strong presence of one color. Later, in Chapters 4 and 5, you will see the mix of land covers present in each bar of the graph become increasingly evenly distributed. The final patch type in Chapter 5 is an evenly heterogeneous patch with equal mixtures of the five land covers: woody vegetation (dark red), herbaceous vegetation (light red), bare soil (green), pavement (light blue), and buildings (dark blue).

coinciding with the Soldiers Delight conservation area is located on unusual, infertile soils along the western border of the watershed. Both these instances of large woody patches are associated with land that was inhospitable to row crop agriculture, or inimical to development because of steep slopes.

The forests present along streams outside of dense urban areas have been shown to help reduce nitrate pollution in stream waters. In contrast, research in inner-city Baltimore has shown that streamside woody vegetation does not have the expected benefit of controlling nitrate pollution in streams. This finding has been used to develop a policy that seeks to spread the benefits of tree canopy throughout the city by planting more trees. Stormwater flows are expected to be mitigated by this policy of dispersing the canopy as well.

Many of the forest-predominated patches in the upper Gwynns Falls watershed are inhabited. The lower watershed relegated forests as public parks. Many suburbs are built on formerly herbaceous farmland, and over time have become increasingly woody. The social implications of forests are many and diverse, depending on the demographic and cultural contrasts across the watershed and over time. While many people find spiritual or aesthetic nourishment under the trees, for others it can be a place of isolation and fear.

Herbaceous Vegetation
Patches that are predominantly herbaceous vegetation have several origins. Some can consist of frequently mown grass such as lawns, golf courses, and cemeteries, while others support lightly managed grass and volunteer broadleaved herbs. Such patches

exist throughout the watershed and there is great variation in the size of the patches. There would have been few native grassland patches in the Baltimore region on upland sites, although herbaceous wetlands would have been important in lowlands. There are positive and negative ecological and social consequences related to the management activities needed to maintain these patches. Typically, large patches of grass are treated with fertilizers and herbicides, which may affect soil health and stream water quality. In addition, mowing to maintain grass as lawn incurs costs of labor, fuel, and air and noise pollution.

The green lawn is of great cultural importance in the United States, creating visual continuity between separate properties, and also marking civic spaces such as public parks, recreation fields, cemeteries, and golf courses. Farms within the watershed are also considered herbaceous. For some, these patches can be aesthetically pleasing in their homogeneity as great care is often taken to remove unwanted plants from the mix. But this loss of biodiversity from the urban system may impact communities of pollinators and other organisms that rely on a diverse community of plants. In contrast, vacant lots generally support volunteer vegetation, and are rarely mown, so diversity can be quite high, as the volunteer plant community may include species that disperse in from the surrounding area as well as species that were intentionally planted by previous occupants of the lot. These spaces may also be mown to give the impression of attention or to neaten the streetscape.

Bare Soil
Patches that are predominantly bare soil are rare in the watershed, because the Gwynns

Falls watershed is both naturally verdant and largely developed. These patches are often areas that are neglected or in transition, meaning that the previous land cover has been removed and the vegetation scraped in anticipation of a new use. Some transportation or utility rights of way, such as railroad tracks, also are predominated by bare soil to minimize management of vegetation or reduce vegetative interferences with infrastructure.

Patches predominated by bare soil exist throughout the watershed. They are of variable sizes but most are relatively small, particularly in the lower part of the watershed. In these areas, the bare soil patches may be the forgotten spaces that awkwardly lie between more maintained or intensively used spaces, such as areas that serve as distribution centers for truck transport or are legacies of past light industry. In the upper watershed, the bare patches are larger and are likely areas of new development. These patches tend to also contain pavement and herbaceous vegetation, as roads may be established first before structures are built.

The bare soil that characterizes these patches infiltrates water on site but is also susceptible to erosion, potentially increasing dust exposure to nearby residents, and is likely to become very compacted and sealed as the patch transitions to one with more buildings and pavement. The social impact of these patches varies depending on how the bare soil is perceived—such as a derelict space that needs attention or as a blank slate suggesting endless possibilities.

Pavement
Patches predominated by pavement are either major roadways or parking lots. These patches are relatively rare because the minimum mapping size of 20 × 20 meters in two orthogonal directions prevents a single road, such as a residential street, from qualifying as its own patch. Instead, the pavement contained in smaller roads or parking lots is accounted for among the mix of land cover elements found in more heterogeneous patches described in later chapters. Many of the patches predominated by pavement are linear and are found primarily in the bottom two-thirds of the watershed. These patches often reflect the transportation infrastructure of postwar suburbanization, or in some cases retrofitting the city transportation corridors to better accommodate the private automobile. A few large pavement-predominated patches are retail malls or warehouses beyond the urban center. The historical urban center, to the extent that it exists in the Gwynns Falls watershed, had relatively few instances of patches dominated primarily by pavement.

While large swaths of pavement are used to move people and products, these patches predominated by pavement do come at an ecological and social cost. Impervious surfaces speed the discharge of rainwater, causing stream erosion and flooding. Large roads moving lots of traffic quickly through the city may physically divide neighborhoods, making them unfriendly to pedestrians and isolating residents from amenities that become difficult to reach on foot. It may also result in noisy streetscapes that discourage casual interactions and engagement with neighbors and small businesses. Traffic noise may also interfere with organisms like birds that rely on auditory cues to communicate with potential mates or guard territory. In many urban situations birds have altered the volume and frequency

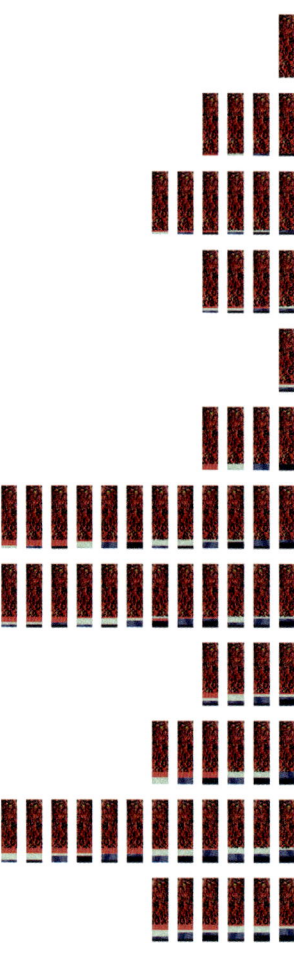

FIGURE 8 Woody vegetation. From left to right, this figure shows a) suite of possible patch types, b) suite of actual patch types, c) location of actual patches of this type in the Gwynns Falls watershed, and d) an illustrative patch of this type. A similar sequence of images is repeated in Chapters 3, 4, and 5. How each of the actual patches dominated by woody vegetation are found in relation to the full watershed array and the region are shown in figure 1. How the suite of patches dominated by woody vegetation fit into the full periodic table of possible patch types is shown in figure 2, and how the suite fits into the full graph of actual patch types in figure 3. A side-by-side comparison of the full periodic table of possible patch types and the full graph of actual patch types appears at the end of Chapter 6.

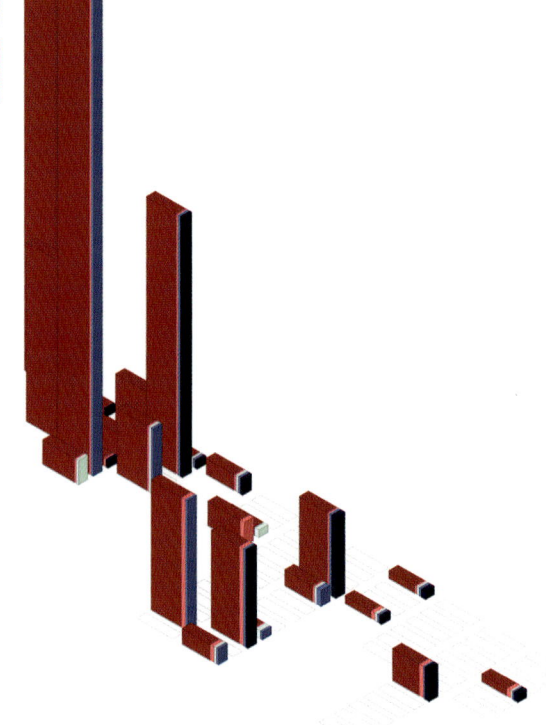

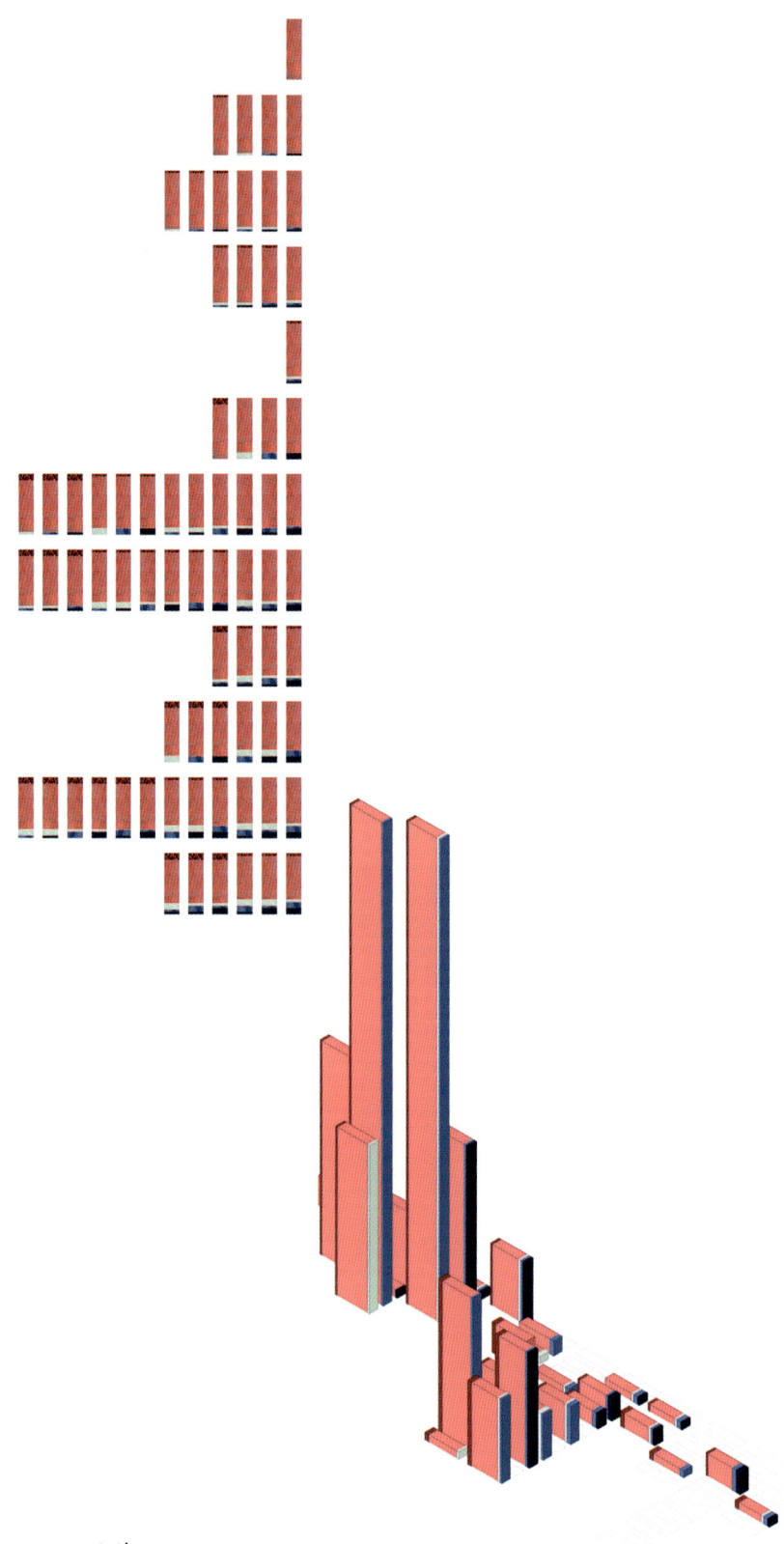

FIGURE 9 Herbaceous vegetation.

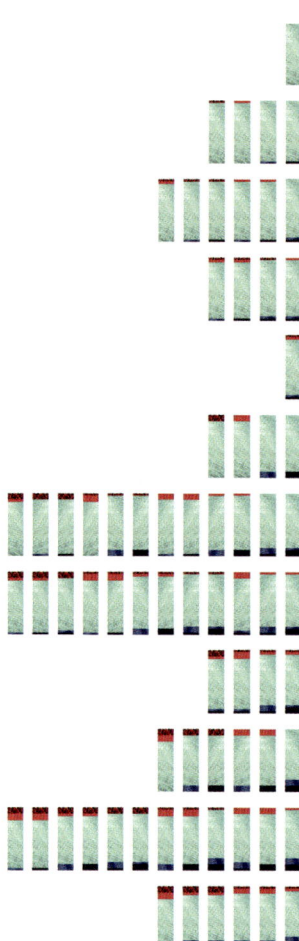

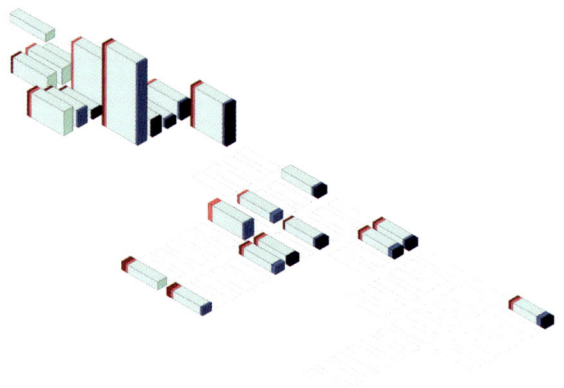

FIGURE 10 Bare soil.

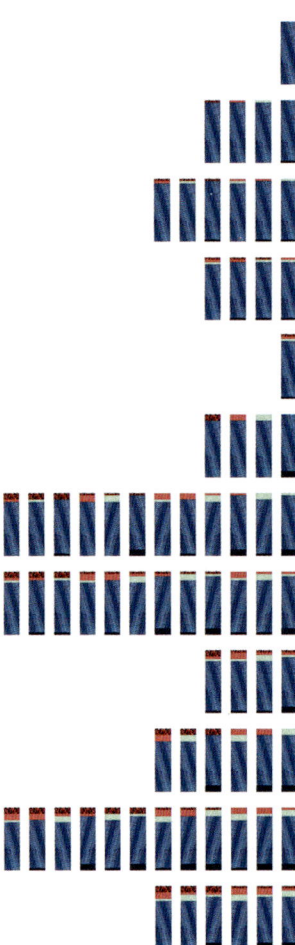

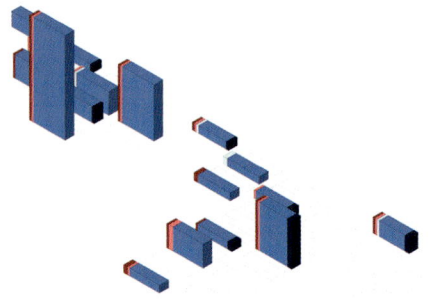

FIGURE 11 Pavement.

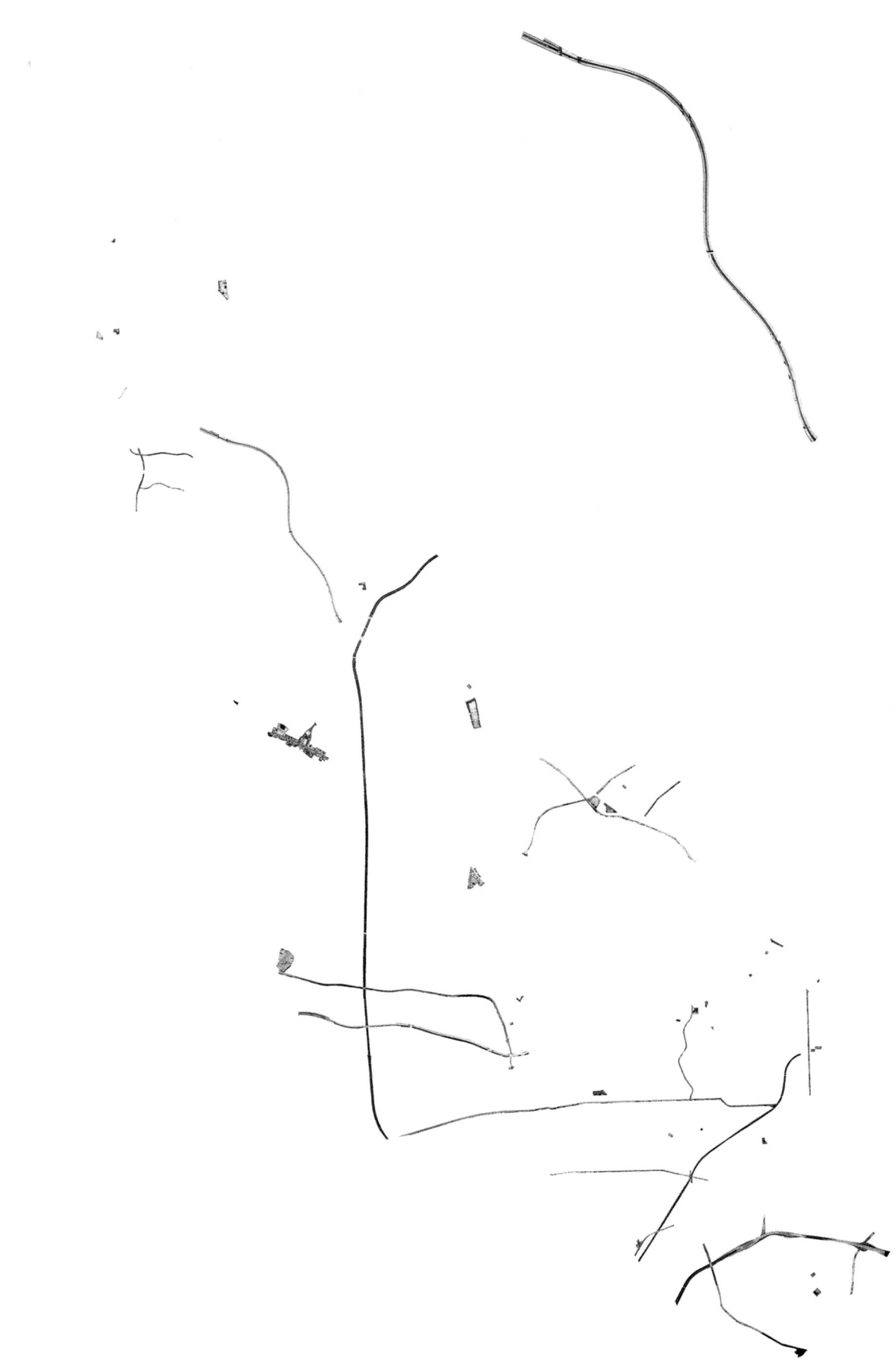

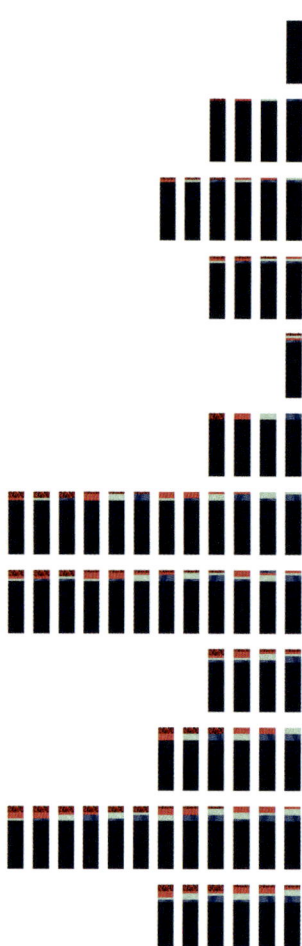

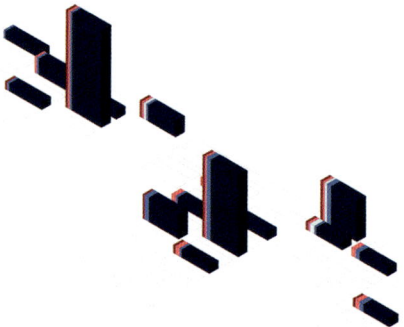

FIGURE 12 Buildings.

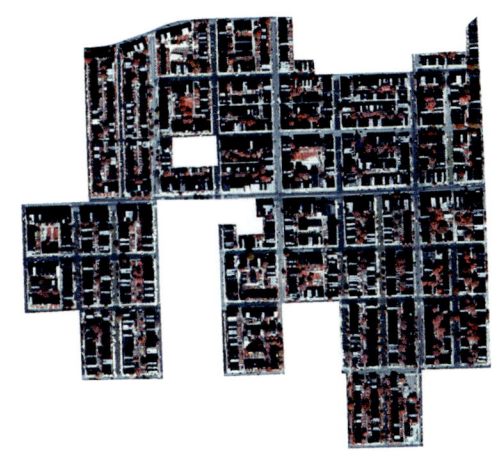

a   b   c   d   e

FIGURE 13 Patch sample contact sheet.
a. Woody vegetation b. Herbaceous vegetation c. Bare soil d. Pavement e. Buildings

of their song in an effort to be heard over urban noise.

Buildings

Despite the Gwynns Falls watershed being highly constructed, few patches are characterized as more than 75 percent occupied by buildings. In general, buildings tend to be accompanied by other land cover elements such as pavement or vegetation, and, in fact, it is rare to find a neighborhood without tree cover or lawns. However, patches that are predominated by buildings are generally found in the lower watershed or the upper watershed, whereas these patch types are conspicuously absent in the middle of the watershed. In the lower watershed, historic rowhouse neighborhoods constitute many of these patches. In the upper watershed, patches characterized by large buildings that are not associated with paved parking lots represent heavy industry, and

are often located along the rail lines. In the middle of the watershed, early twentieth-century suburban residential and commercial development resulted in buildings being closely associated with pavement and vegetation, which occupy more than 25 percent of these patches.

The rowhouse-dominated neighborhoods that comprise many of these patch types in the lower watershed are neighborhoods where residents gather on the front stoops to chat with neighbors, to watch children at play in the streets, and to cool off in the evenings. In contrast, the back alleys, which were intended not for social interaction but rather for the storage of cars and garbage, are rarely used. Because buildings occupy such a great proportion of the patch, these patches both absorb and generate heat, and for both reasons this patch type may experience greater temperature extremes during the hot summer months.

CHAPTER 3 HETEROGENEITY AS OUTCOME OF URBAN TRANSFORMATION: ONE PREDOMINANT LAND COVER ELEMENT WITH GREATER POTENTIAL FOR MIXTURE

This chapter focuses on an area of the periodic table that has a greater potential for mixture of land covers than the patches with one predominant land cover discussed in the previous chapter. The greater potential heterogeneity of this region of the periodic table includes a much greater number of actual patches than those appearing in the previous chapter. The number of patches reflecting predominance of 50–75 percent occupies a greater range of land cover elements—that is, the horizontal dimension of the periodic table, and a greater number of the categories of predominance that appear on the vertical axis in the periodic table. Thus, both the breadth and depth of the chart of actual patches are more fully occupied by the patches discussed in this chapter than those in Chapter 2. The arrays of potential and actual patches result from different logics.

The distribution of patch types discussed in this chapter is due in part to the logic of combinations that this portion of the periodic table lays out. Slightly predominant patches have a greater possibility of potential mixture. In this part of the periodic table, one HERCULES land cover element occupies between 56 and 75 percent of the patch, leaving 25 to 44 percent of the patch to accommodate the other four cover elements. Thus, there is a rich array of mixtures that this portion of the periodic table can exhibit. However, the logic of combinations doesn't explain what actual patches and their abundances will be.

The mixture of patches present in Baltimore depends on a second logic, which emerges from urban transformations over the twentieth century. The century spanned shifts from the compact industrial city to the engineered sanitary city, from fine-scale racial, commercial, and industrial adjacencies to coarser-scale regional heterogeneity and segregation, from few and small green spaces, to regionally designated green infrastructure, from reliance on coal and rail to use of gasoline and the private automobile, and from the progressive-era governance for the public good to neoliberal private/public partnerships. These shifts mean that new spatial heterogeneities have been emerging over the course of the twentieth century at various scales. In addition, there are legacies

FIGURE 14 An example of land that is predominantly covered with herbaceous vegetation. Other land covers are also present, including buildings, pavement, and woody vegetation.

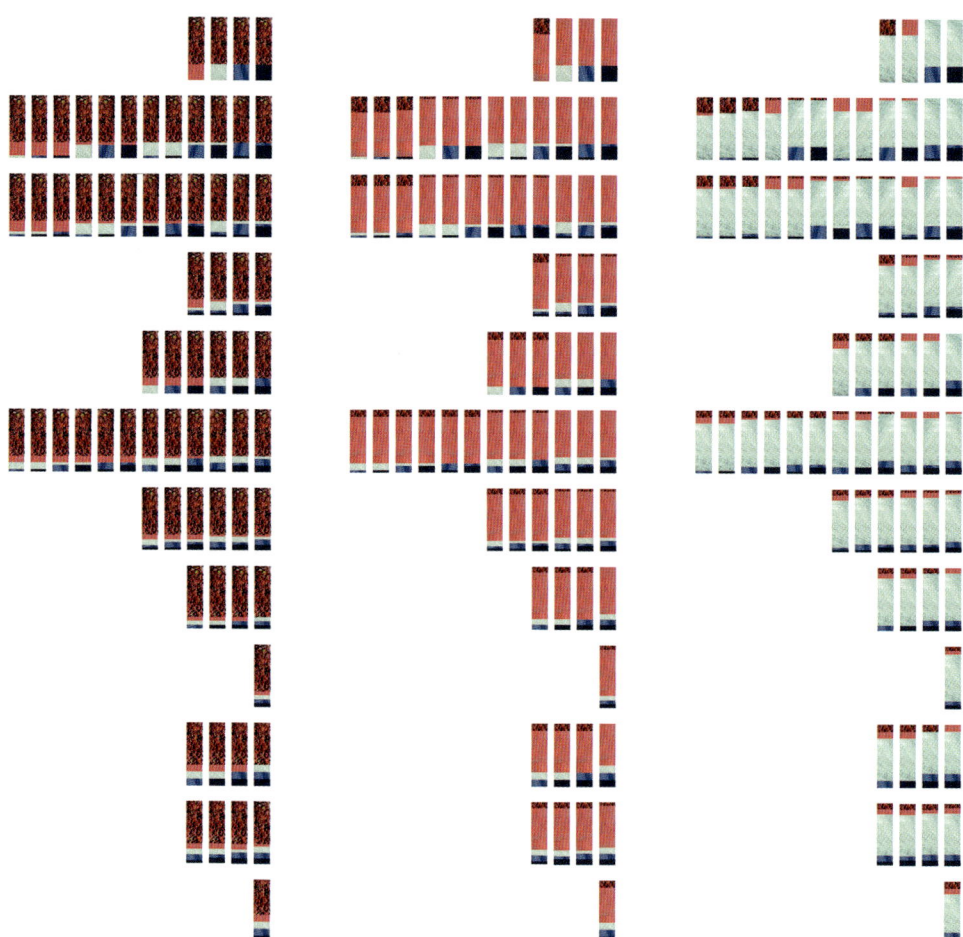

FIGURE 15 Overview, 75–49 percent.

of the fine-scale, rowhouse, and industrial city patch types that have been altered, and new patch mixtures that have been designed or have simply evolved in relation to the new economic and governance strategies. Thus, there are many causes of the diversity of actual patch types.

The most pervasive cause was the spread of the Olmstedian suburban ideal. This picturesque view of city life, having private backyards and shared greensward in front, became a norm. This ideal drives the mixture of land cover elements described in this chapter. Because this shift led to a dependence on the private automobile, paved sur-

faces become more extensive. The mixtures also reflect the dispersion of commercial, logistical, and residential densities throughout the region, not just in the core city. This is the urban form most conducive to normative landscape analysis, where patterns of remnant green space are prioritized, but it ignores the built fabric and gray, or constructed, infrastructure of this more dispersed city.

The suburbanization of the watershed was associated with the national shift from small farms to petroleum-powered agribusiness, which opened peri-urban land to suburban development. The shift was also

favored by subsidies for interstate highways, inner city redlining, federally guaranteed suburban mortgages, and large-lot zoning. Commercial aggregation changed from small businesses embedded in the urban fabric to large, automobile-oriented strips and indoor malls. Accelerated globalization shifted manufacturing overseas, driving industrial and residential abandonment in parts of the watershed, and a sprawling trucking network of distribution and "fulfillment" centers.

While many people benefitted from the trends in twentieth-century urbanism, others have been marginalized and bypassed. The century generated an "urbanism of exclusion," enforced by a variety of legal, informal, and interlocking policies and practices. Although some of these exclusionary practices have been replaced, their legacy remains, and has even been exaggerated as wealth disparities have dramatically increased in the United States since the mid-1970s.

Woody Vegetation Moderately Predominant
This patch suite reflects the integration of woody vegetative and built elements, expressing the suburban aesthetic. There are two large clusters of this patch suite in the

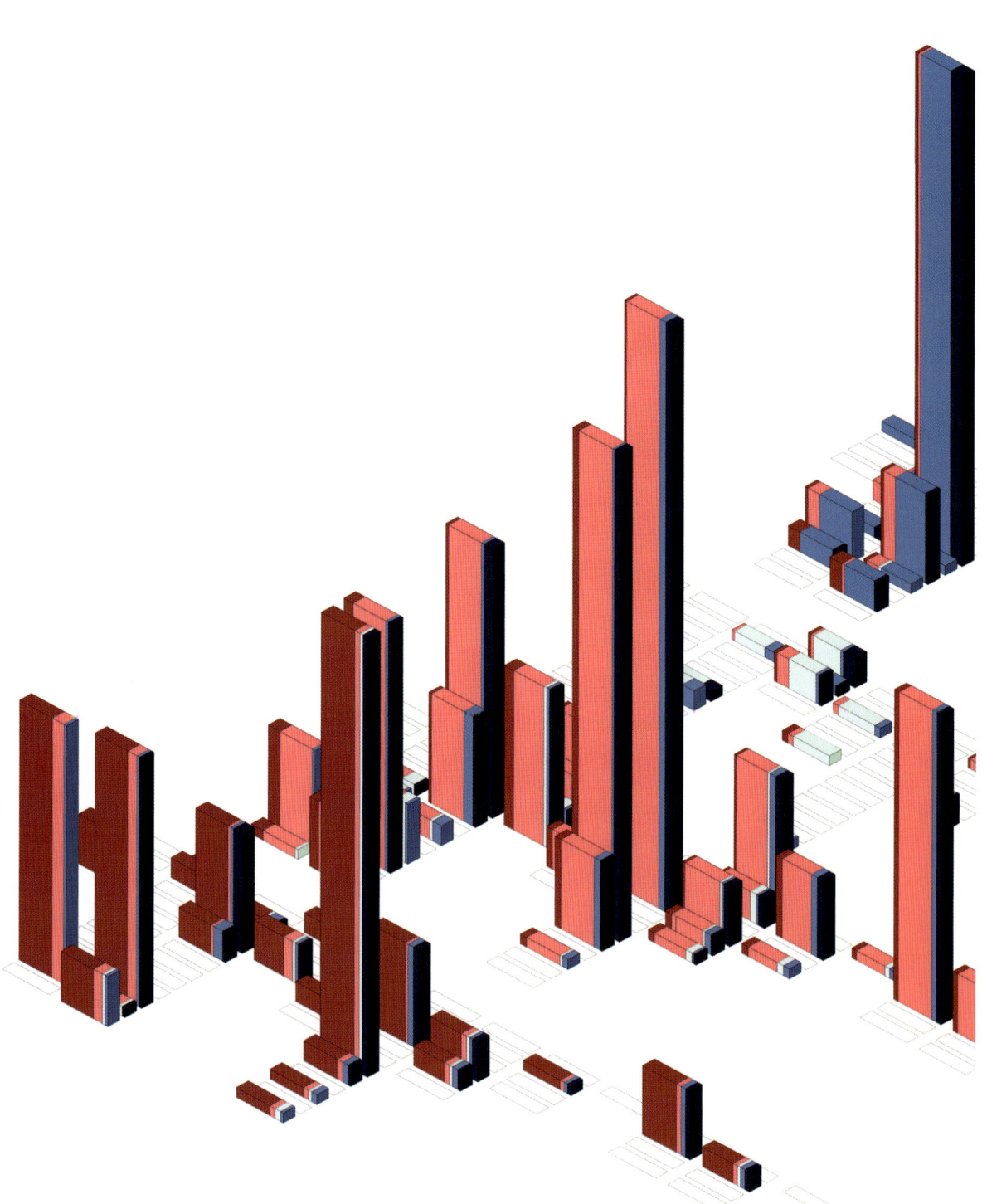

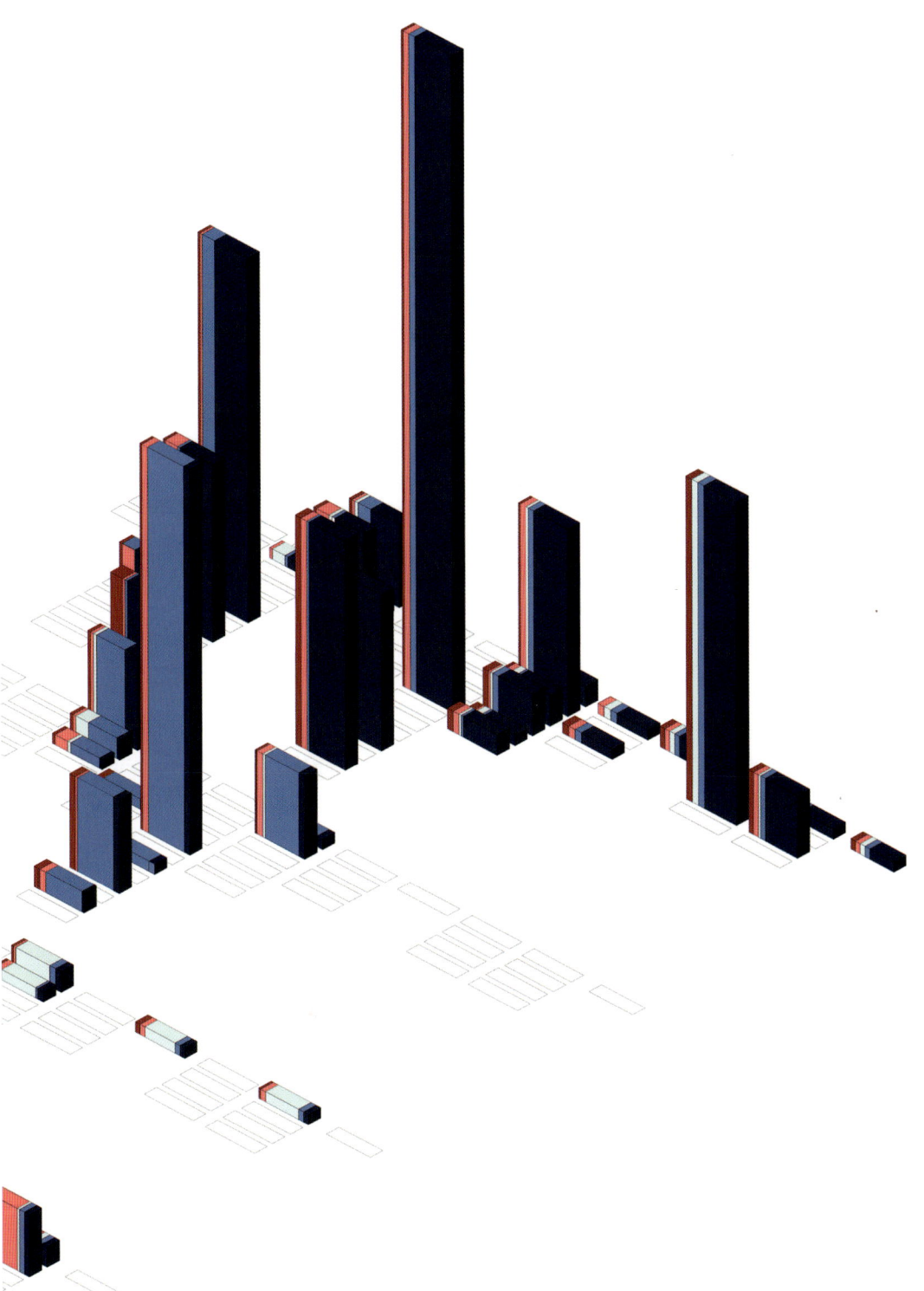

FIGURE 16 Overview, 75–49 percent.

watershed. These are older neighborhoods—"streetcar suburbs"—that were planned and purposefully managed to support trees. These patches include neighborhoods adjacent to the large, forested Gwynns Falls and Leakin Parks. The benefits of large clusters of moderately dominant forest patches are expected to be locally focused because these effects decline with distance. It is important to recognize that the lack of tree canopy in other neighborhoods may reflect the aesthetic choices of earlier generations of residents, or the wariness of current residents about the burdens of management that trees may place on them. For example, the lack of municipal care of trees in underserved neighborhoods may contribute to negative views of tree canopy by residents there.

The tree canopy within the watershed is highly fragmented, limiting its ecological functions. Fragmentation can constrain bird diversity, for example. In spite of such constraints, patches moderately dominated by trees provide important habitat for both resident birds and seasonal migrants. A variety of other organisms, ranging from coyotes to small rodents, but also including inconspicuous soil invertebrates important in nutrient recycling, rely on relatively complete tree cover. In addition to providing species habitat, these patches may ameliorate the local climate, helping to offset the urban heat island effect and improve the thermal comfort of people. Predominant forest patches also contribute to stormwater management and to recreational benefits.

Herbaceous Vegetation Moderately Predominant

Of the seventy possible patch types characterized by having moderately dominant grass cover, most of them actually appear in the Gwynns Falls watershed. Furthermore, patches of this suite are distributed relatively evenly throughout the watershed. This patch suite represents large agricultural fields, new suburbs, or suburbs with little tree cover that may become dominated by mature trees over time. Patches of this type also appear along managed highway rights-of-way.

The moderate predominance of herbaceous vegetation may exist for many reasons. Some may be temporary, as in the case of new residential development, or lands long released from agriculture but not yet subdivided. Some may be permanent, as in the case of septic fields, recreational fields, institutional grounds, and residential mown turf. But it is the cultural preference for the maintained and manicured lawn that seems to ensure the moderate predominance of herbaceous vegetation.

Lawn is often considered to have no positive ecological function; however, this land cover element has been associated with a surprising amount of ecological activity. Although there are fuel and pollution costs associated with intensive turf-grass management, there are some ecological processes that are facilitated in parts of the urban ecosystem supporting grass covers. In particular, appropriately managed lawn can help reduce nitrate pollution in ground and stream water, as well as contribute to carbon storage.

Bare Soil Moderately Predominant

This patch suite consists of patches with moderately predominant bare soil, and these are represented by remarkably few actual patches in the Gwynns Falls watershed. Those that are present are evenly distributed throughout the watershed. Bare soil is likely a land cover element in flux, and is found to be present mainly in areas of new development. An exception is the bare soil of baseball infields, which is required by the game. This scarcity reflects the cultural predisposition for vegetative growth in non-paved open space, as well as a climate conducive to growth of unirrigated trees. A dry, desert city would likely have a much more common presence of bare soil.

Ecologically, bare soil may be disturbed, may present a compacted and relatively impervious surface, or may be material brought in as fill from other locations. Fill may develop into a well-structured soil over time, but it is a slow process. In general, bare soil, with its chemical and physical properties affected by demolition, construction, and movement of materials, can influence the cycling of nutrients and the distribution of inorganic and organic contaminants. Although bare soil is normally present during plowing cycles of agricultural land or seeding in lawns, these are considered as herbaceous vegetation.

An example of the social-ecological significance of bare soil is the presence of lead in any exposed soil in residential yards. Although the traditional concern with lead as a hazard to urban residents, especially children, has focused on lead paint in building interiors, the contamination of soils outside the home is now widely recognized as a serious threat. Lead in soil is a legacy of leaded gasoline, and of lead-based paints. Although the United States banned lead paint in 1978, and leaded gasoline in 1996, lead contamination in soil persists for very long times. Lead is especially problematic near building walls, and along heavily traveled roads. Children can be exposed to lead while playing in areas of bare soil in residential yards.

Pavement Moderately Predominant
Actual patches in this suite exhibit more isodiametric shapes than the patches supporting more than 75 percent pavement described in Chapter 2. Many of these patches are artifacts of the interstate era, with its growth and spatial spread of the "car culture" that emerged when streetcars of the Olmstedian suburbs were abandoned. Parking areas for big-box stores and indoor malls contribute to this patch suite. Other kinds of paved open space, such as public plazas or private courtyards that appear in many Latin American, European, or ancient Asian cities, are rare in the Gwynns Falls watershed, where public open space and private gardens are most often vegetated.

Pavement efficiently absorbs energy from the sun, warming up during the day and slowly reradiating stored solar energy once the sun goes down. This, in part, gives rise to the urban heat island effect, whereby cities in moist climates that have large amounts of paved areas are warmer than their surrounding rural or vegetated areas. The heat reradiated by pavement after sunset prevents pavement-dominated areas from cooling down after a hot day, and patches having a lot of pavement are less thermally comfortable for residents.

Pavement-dominated patches are also major contributors to urban stormwater runoff. Runoff from paved surfaces is intentionally rapid, which is a major contributor to the flashy nature of urban streams—flood waters rise rapidly, and the resultant stream flows are deeper than in streams not connected to paved areas. Contamination by oil, de-icing salts, and heavy metals are other

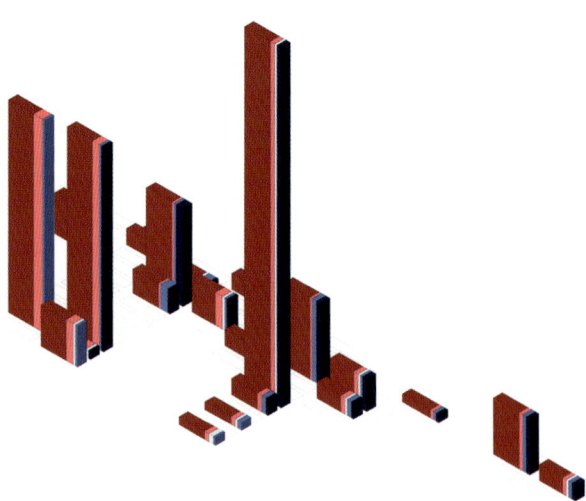

FIGURE 17 Woody vegetation +.

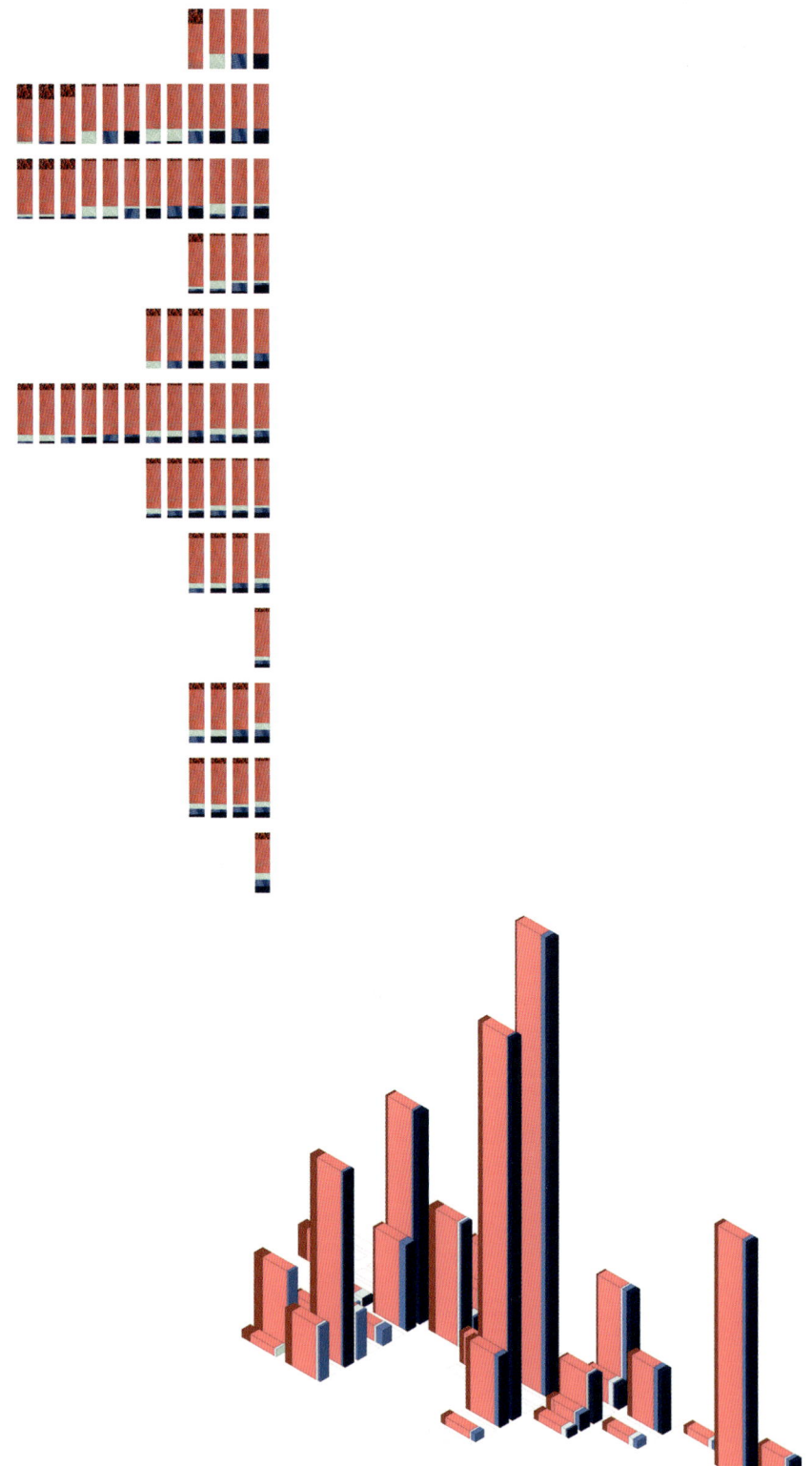

FIGURE 18 Herbaceous vegetation +.

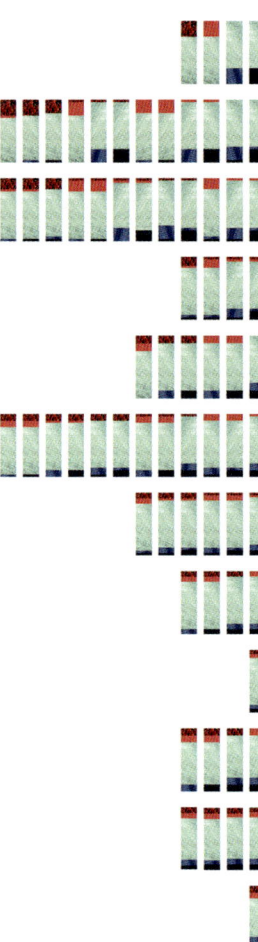

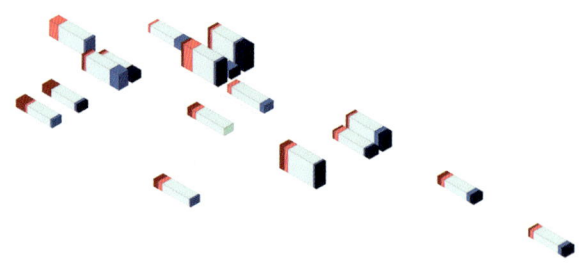

FIGURE 19 Bare soil +.

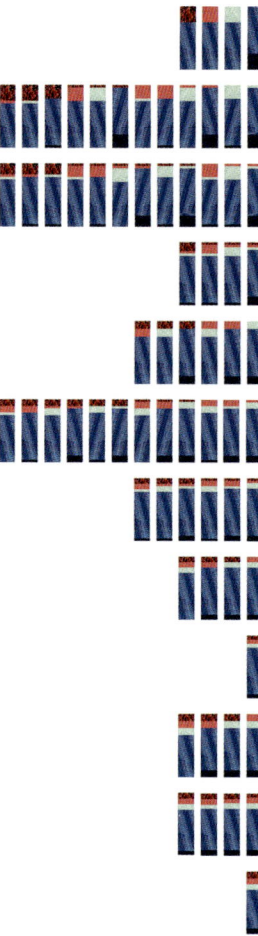

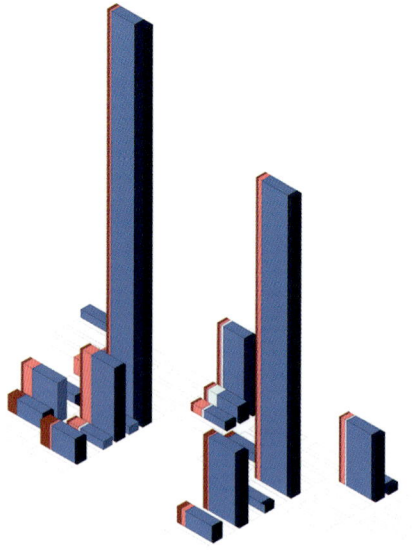

FIGURE 20 Pavement +.

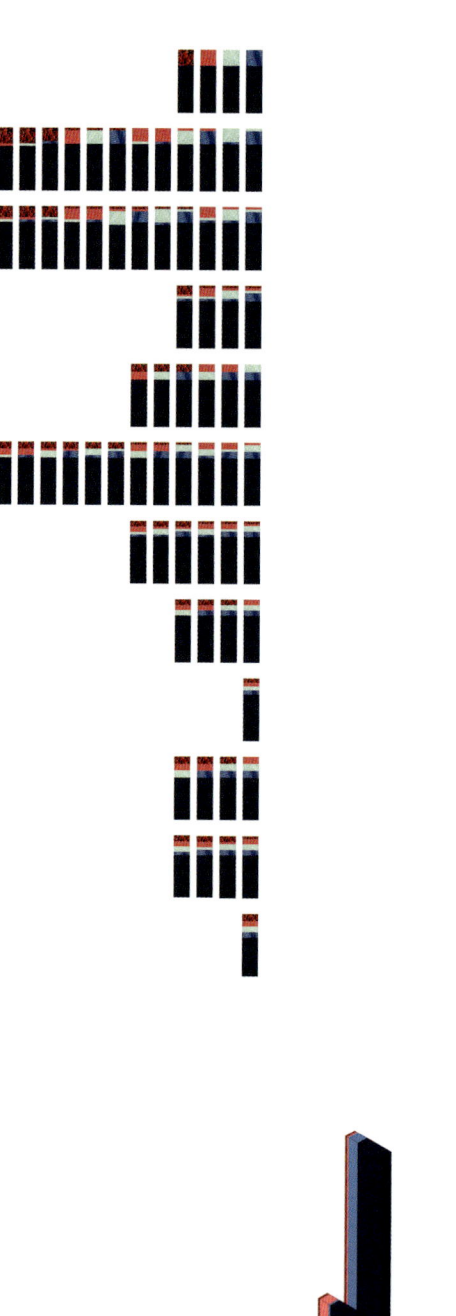
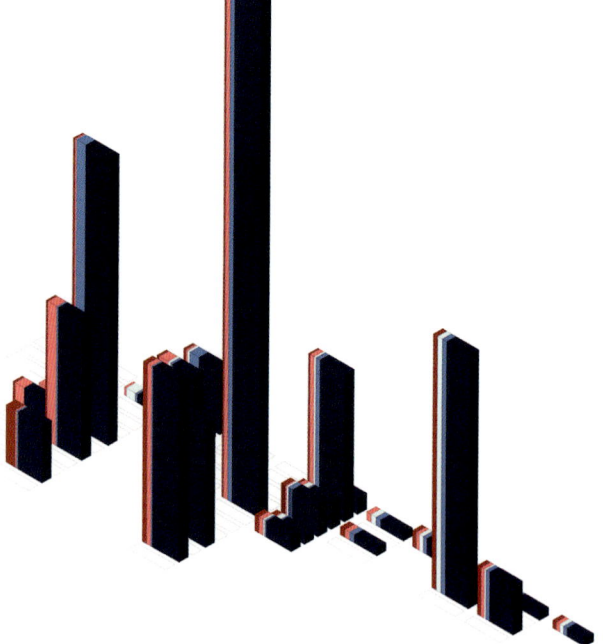

FIGURE 21 Buildings +.

FIGURE 22 Patch sample contact sheet.
a. Woody vegetation + b. Herbaceous vegetation + c. Bare soil + d. Pavement + e. Buildings +

effects of paved areas. Finally, the heat accumulated by asphalt and concrete pavement not only increases air temperatures, but it also increases the temperature of urban streams, because storm runoff picks up heat as it runs across paved surfaces.

Buildings Moderately Predominant
This patch suite of cover types has many different mixtures of land cover elements along with buildings, so that its frequency distribution in the chart of actual covers extends through the vertical dimension—this is called deep, meaning there are a lot of these types of patches present in the watershed. This patch suite is also widely distributed throughout the watershed. For example, patches with buildings moderately dominant exist in leafy rowhouse neighborhoods, and in new high-density apartment and condominium developments along the beltway, and in the new, outer suburbs. These patches have a high density of build-

ings and, importantly, are not restricted to the old city areas of the watershed. However, the color shift from blue to red moving up the watershed in this suite of moderate dominance of buildings indicates that there is a greater presence of vegetation within these moderately predominant building patches as one moves northwest up the Gwynns Falls watershed.

The broad distribution of patches in this suite is evidence that a simplistic expectation of linear decline in density or in urban effect as one travels outward from the old core is untenable. Although urban gradients are important tools for research and explanation of urban change, such gradients must be abstracted by calculation and statistics from any linear transects through an urban regional mosaic. A typical transect along or across the Gwynns Falls watershed would exhibit an irregular distribution of patches rather than a linear change in patch features.

CHAPTER 4 REGULARITY WITHIN PATCHES AS A CHARACTERISTIC OF HETEROGENEITY: TWO CO-DOMINANT LAND COVER ELEMENTS AND REPEATED PAIRS

This suite contains patches that have two predominant land cover elements. The percentage of each predominant element is 27–50 percent, and therefore these patches can be thought of as the middle ground of the periodic table. They are the transition from those patches that are dominated by one land cover element toward those patches that have a mix with all elements equally present.

In the periodic table there are ten possible combinations of co-dominant land cover elements within patches. The five most common pairs have been selected for illustration: woody vegetation and herbaceous vegetation, woody vegetation and buildings, herbaceous vegetation and pavement, bare soil and pavement, and pavement and buildings. Such pairings represent the land covers that tend to be co-present in a patch along Gwynns Falls, but they also reveal pairs that do not exist, such as her-

baceous vegetation and buildings, or woody vegetation and pavement.

In the chart of actual patches, there are two land cover combinations that are abundant as repeated pairs. One pair is woody vegetation and herbaceous vegetation (seen on the left-hand side of the figure), and the other pair is pavement and buildings (on the right-hand side of the figure). There are more repeated pairs at extremes of the array of actual patches that are less abundant. You can also see that patch types representing paired co-dominance are distributed across the vertical or depth dimension within each suite, meaning that they can be seen from top to bottom of the suite.

Whereas the earlier chapters unraveled the concept of homogeneity and the way that urbanization shapes heterogeneity, this chapter looks carefully at five pairs in order to explain the characteristics of regularity. Pairings are important because different land cover elements have different ecological structures and functions, and when these are regularly found in combination, it may be productive to use them to guide research and management actions or policies. Because one of the outcomes of urban design is the intentional combination of various land cover elements, it may be

FIGURE 23 An example of land that has two predominant land cover elements: in this case, pavement and buildings, along with a small amount of herbaceous vegetation. The image also includes a patch boundary along the edge of the parking lot: to the left is land dominated by woody vegetation.

FIGURE 24 Overview, 50–36 percent.

said to have a combinatory logic. Therefore, urban design may benefit from an understanding of the regularity of certain pairings within an urban system.

A sharp contrast between land cover elements is where there is structural discontinuity or distinction. A harsh contrast is one where there are functional consequences or effects. Sharp and harsh characteristics of regularities can occur between elements within a patch, as well as across the boundaries of a patch—and in relation to the mix of elements in that adjacent patch. There can be subtle contrast, and this can often be found in narrow or unusual patch shapes.

Finally, there can also be temporal contrast, and this will be explained more in the next chapter, where the characteristic of mutability is introduced.

Woody Vegetation and Herbaceous Vegetation
There are many patches that are co-dominant with woody vegetation and herbaceous vegetation, and they are somewhat evenly distributed across the watershed, with some prevalence in the upper watershed. The shape of the patch sample shown is irregular. The left boundary seems to be partially shaped by a road—an adjacent patch that is highly contrasting. The right boundary

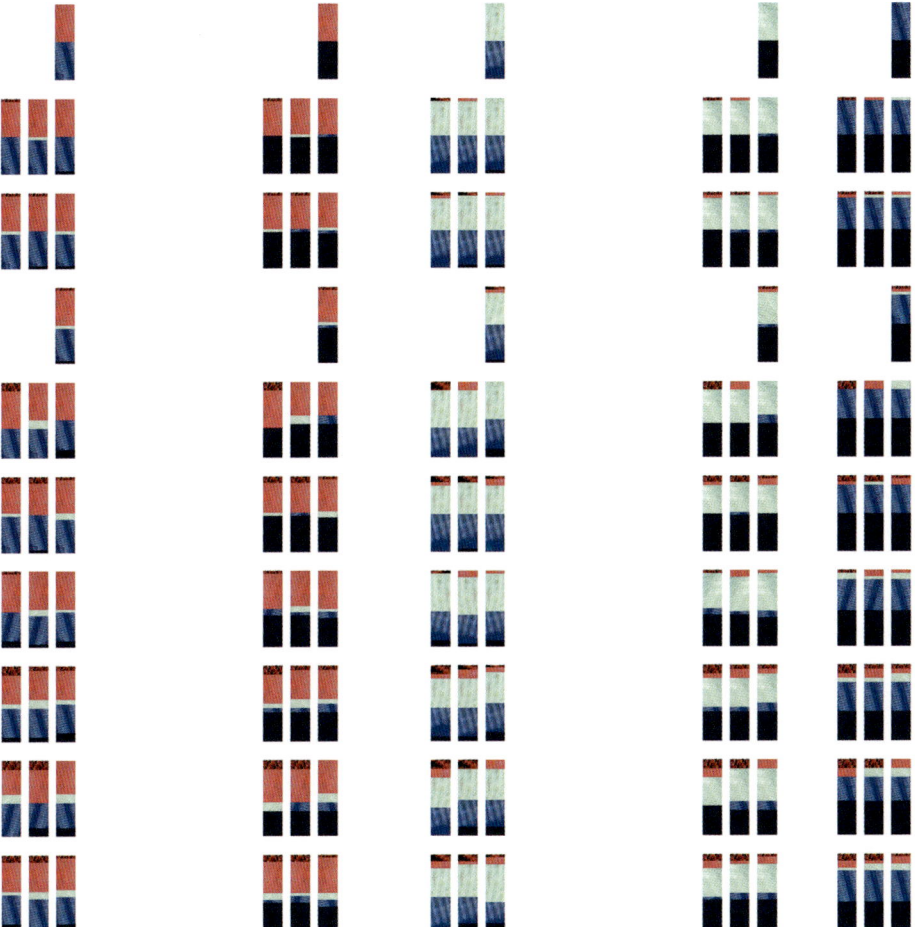

seems to be shaped by an adjacent patch that is subtly contrasting. Another characteristic to note about the shape of this patch, and others similar across the atlas, are the areas where it is narrow. When thinking about flows across patches, these narrow areas may be influenced more by adjacent patches than patches or patch parts that are rounder and or thicker.

The patch sample shown is a residential area that has much grass and trees, as well as some small buildings, roads, and driveways. It is a reflection of the way that patterns of parcelization aggregate and can then be drawn as a patch—in other words,

similar parts of many small land parcels can add up to form a patch that reflects property and ownership patterns. In this case, the patch sample is a combination of new cul-de-sacs with very large plots of land as well as an area with many individual land parcels with driveways onto a village road. These scenic subdivisions often combine lawn and tree clusters as design elements. Other types of patches that form this suite are tree-shaded exit ramps, private or commercial landscaped estates, and large tree-studded park areas with streams and grassy areas for passive recreation crisscrossed by large roads.

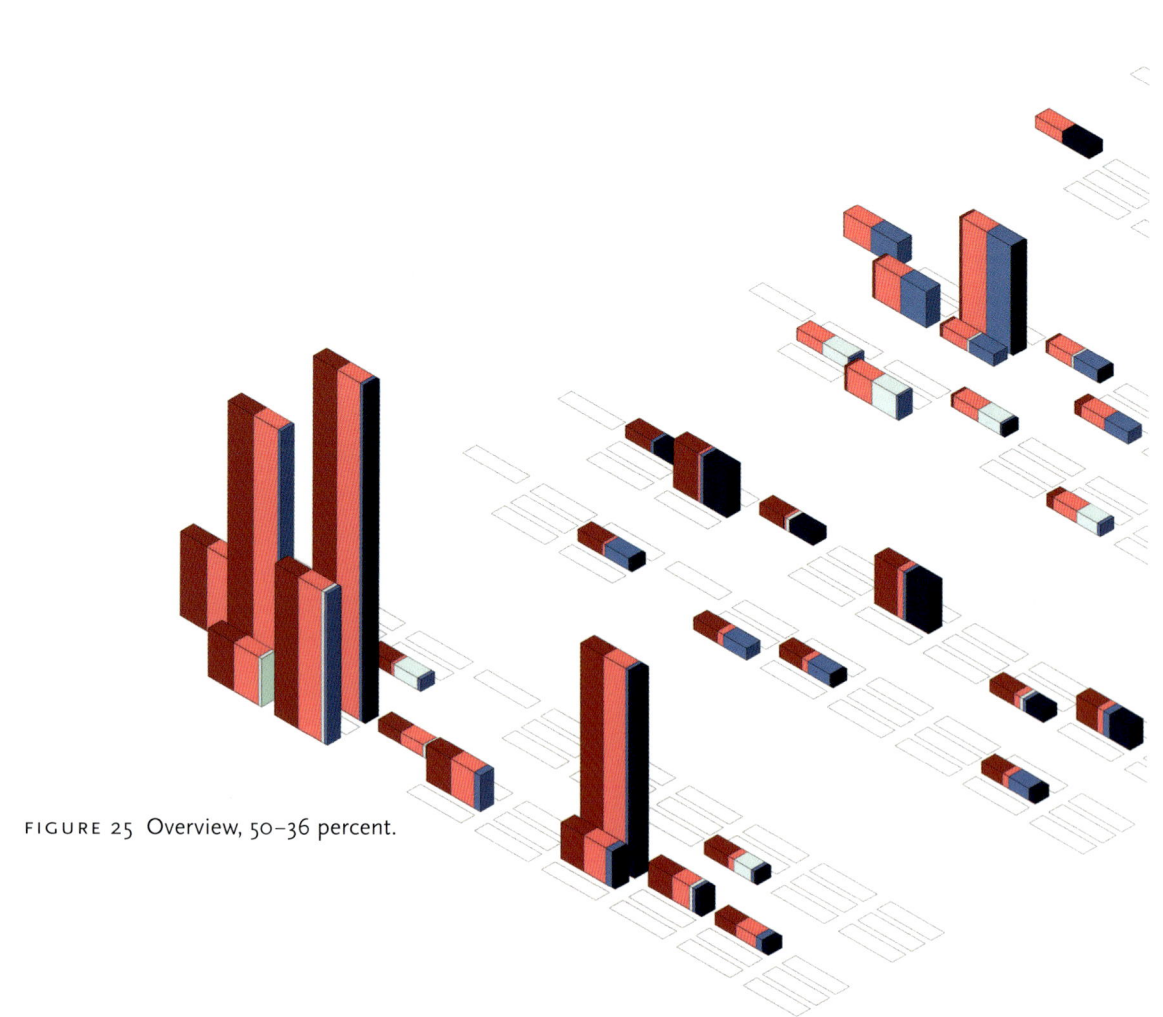

FIGURE 25 Overview, 50–36 percent.

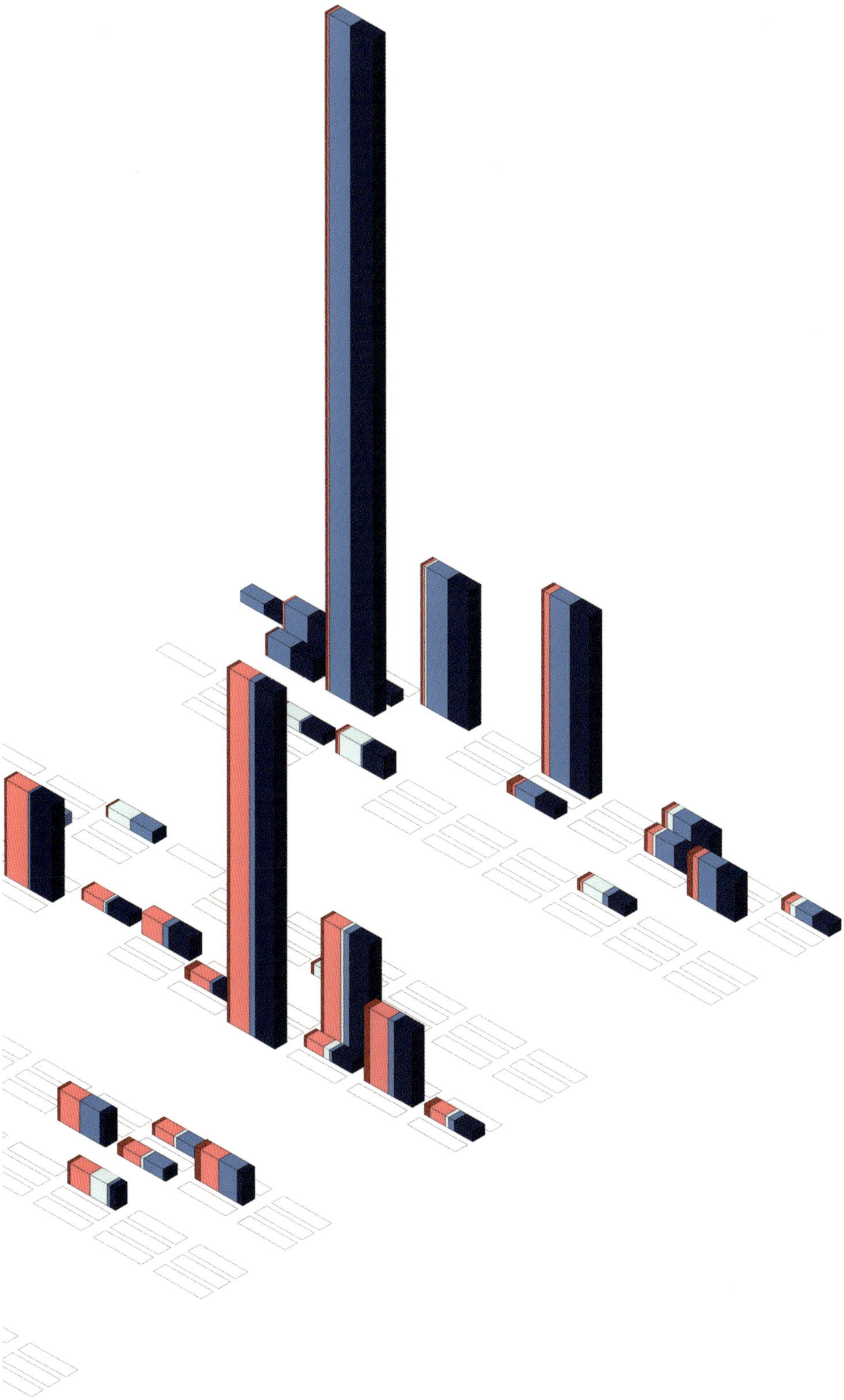

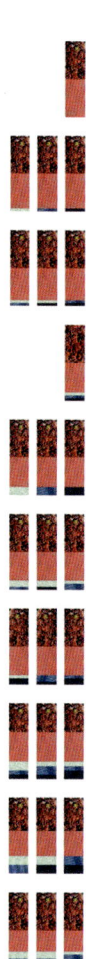

FIGURE 26 Woody vegetation and herbaceous vegetation.

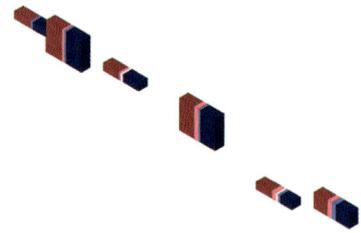

FIGURE 27 Woody vegetation and buildings.

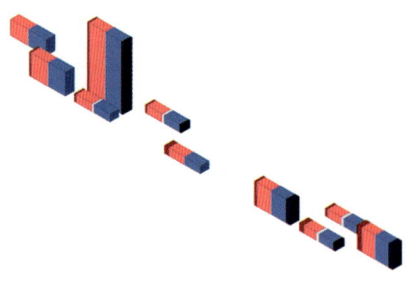

FIGURE 28 Herbaceous vegetation and pavement.

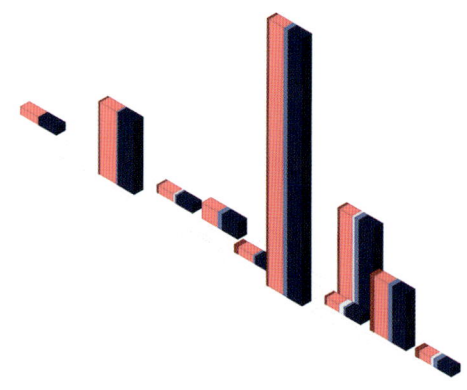

FIGURE 29 Herbaceous vegetation, and buildings.

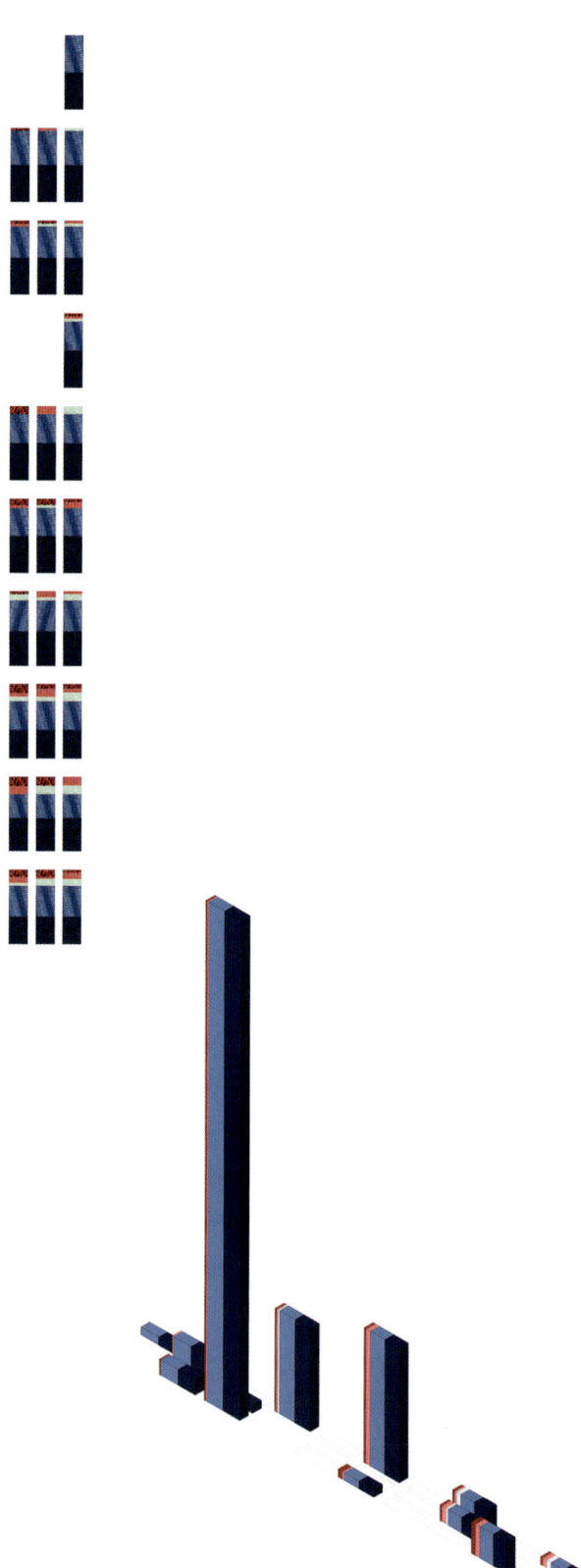

FIGURE 30 Pavement and buildings.

Woody Vegetation and Building

In this suite, there are roughly two types of paired patches; the first are older planned suburbs that are formed by subdivision and parcel accumulation by individual home-owners. This type is found more frequently in the middle and upper regions of the watershed. The second type is older row-house neighborhoods with many vacant parcels. This type is typically found in the lower portion of the watershed and closer to the downtown core of the city. In both types there are continuous and irregularly shaped tree canopies. These canopies may be remnant trees from forests or woodlots that survived in the agricultural landscape, planted street and yard trees, or trees that have spontaneously regenerated in spaces with little to no direct management. In general, there are not many patches in this suite.

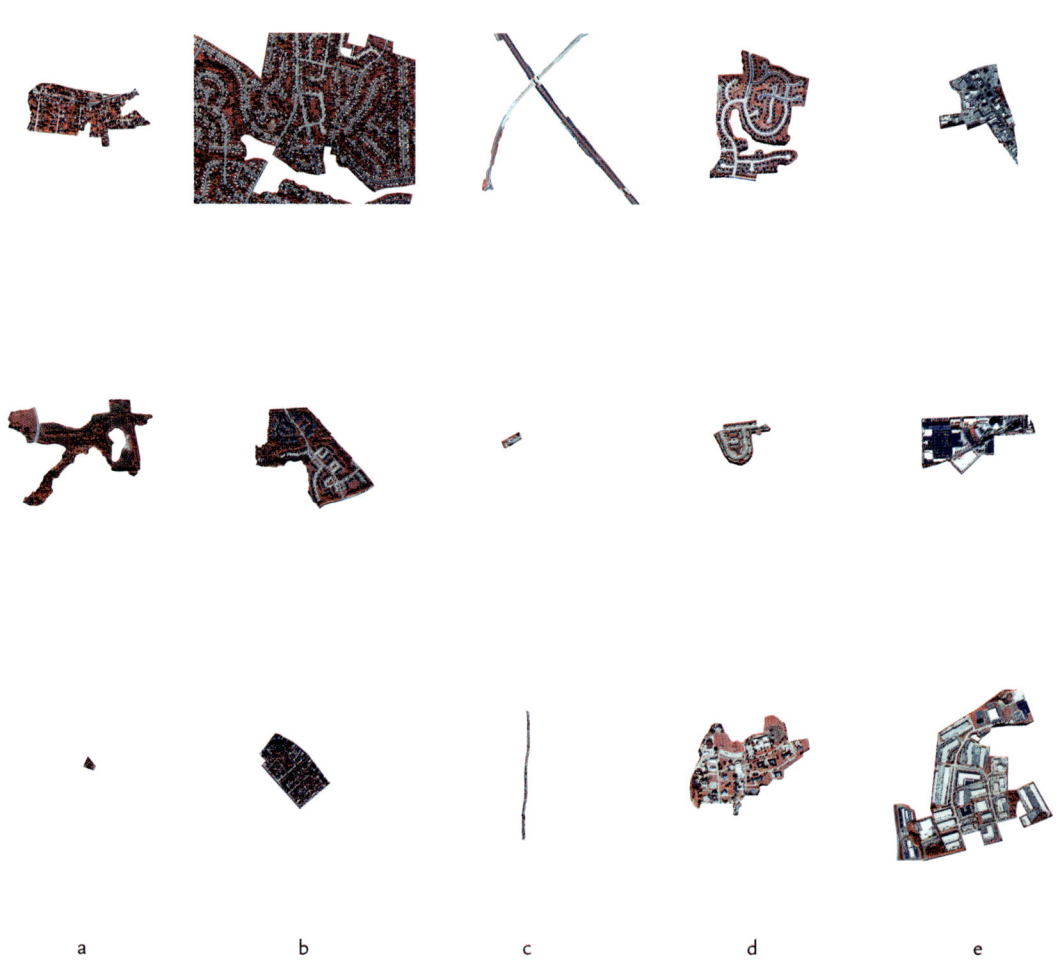

a   b   c   d   e

FIGURE 31 Patch sample contact sheet.
a. Woody vegetation and herbaceous vegetation b. Woody vegetation and buildings c. Herbaceous vegetation and pavement d. Herbaceous vegetation and buildings e. Pavement and buildings

Herbaceous Vegetation and Pavement
Patches that are co-dominant with herbaceous vegetation and pavement are small and scattered across the watershed. They include such structures as a mall with a significant amount of parking, and a suburban church also with much parking. In addition, there are some patches of this type that are very long and skinny. These include a parkway, and the former I-70 corridor referred to as the "Highway to Nowhere" that has wide grass medians. In general, there are few patches in this suite.

Herbaceous Vegetation and Building
Patches that are co-dominant with herbaceous vegetation and buildings are mainly present in the upper and lower watershed and less so in the middle of the watershed. Patches of this type found in the upper watershed are new planned suburbs with ample lawns and very few trees. Patches of this type in the lower watershed are rowhouse neighborhoods. There are many of these patches in the lower watershed, because many rowhouses that fronted on alleys rather than the primary streets were demolished in the 1970s, leaving the inner block spaces to be used for public recreation. In addition, urban renewal projects of the 1970s replaced whole blocks of rowhouses with high-rise, high-density Housing and Urban Development (HUD) projects. These projects have since been replaced by HOPE 6 low-density residences that are suburban in character and contain buildings surrounded by much herbaceous vegetation. Other instances of this patch type are educational campuses, commercial estates, and big-box retail surrounded by mown grass.

Pavement and Buildings
Patches that are co-dominant with pavement and buildings form a distinct pattern. They are arranged as a strip along the western edge the watershed. This is the commercial strip along Reisterstown Road, originally an agricultural market road that connected Baltimore to nearby towns and cities before the construction of the interstate highway network. There are also patches that align to crossroads such as the beltway and highway interchanges, as well as some scattered patches. In typical landscape ecology analysis, these strips are referred to as corridors, or spaces of flows, although in HERCULES these are patches in themselves, and should be addressed as such.

Unlike the patches predominated by pavement, these co-dominant patches have mixed uses. They include industrial and distribution centers serviced by trucks, for example, and many small commercial shops with abundant visitor parking, rather than one big shopping mall that is surrounded by a giant parking lot. Also included in the mix are old village buildings that have been modified from residences to professional offices. The strip represents jobs and consumer society with its infrastructures of shopping, hotels, restaurants, and entertainment. Each building is an air-conditioned box in a little heat island that is shaped by an extensive surface of disconnected parking lots, car traffic, and truck-based goods distribution and waste management. The patches of this type that are scattered tend to be locations of medium-density residential use with on-street parking, and they are typically located near highway interchanges.

CHAPTER 5   THE CASE OF PATCH PLURALITY AS A LESSON FOR URBAN MUTABILITY: THREE TO FIVE CO-DOMINANT LAND COVER ELEMENTS AND THE POTENTIAL FOR RECOMBINATION

This chapter describes patches that have at least three co-dominant elements. These patches are referred to as plural, or having the characteristic of plurality. In this chapter we provide examples of patch suites containing three, four, and five co-dominant land cover elements. We argue that a small change in a co-dominant land cover element may result in a shift in patch identity more readily in this part of the periodic table than in suites characterized by fewer and more dominant land cover elements higher up in the periodic table. Thus, patches co-dominated by three or more land cover elements may exhibit greater mutability than those patches that have one or two dominant land covers.

In the graph of actual patches, there are three suites co-dominated by three land cover elements that are clearly more abundant. These are 1) woody vegetation, herbaceous vegetation, and buildings; 2) woody vegetation, pavement, and buildings; and 3) herbaceous vegetation, pavement, and buildings. Of the five possible suites

FIGURE 32 An example of land that has at least three co-dominant land cover elements: in this case, woody vegetation, herbaceous vegetation, and buildings.

co-dominated by four land cover elements, only one is abundant. This suite contains all land cover elements except bare soil. Few patches exist that are co-dominated by all five land cover elements.

Patterns of more than three co-dominant suites may be due to a number of factors. Patches in these suites appear to be mostly residential and they tend to be large. Their size suggests that these patches reflect large suburban developments that were built in a systematic form combining designed compositions of trees, lawns, and houses. The mixture of vegetation may be determined by the age of development, with older developments having an increased tree cover. The amount of pavement may be associated with the mode of transportation for which these developments were designed and the period of development, as recent planning legislates very wide roads for firefighting equipment. This may be in contrast to other residential patches contained in suites dominated by fewer land cover elements. Those patches may contain residences that were built on larger parcels and contain more vegetation relative to the building or pavement or, alternatively, they may contain residences such as rowhouses that were built on small parcels with little sur-

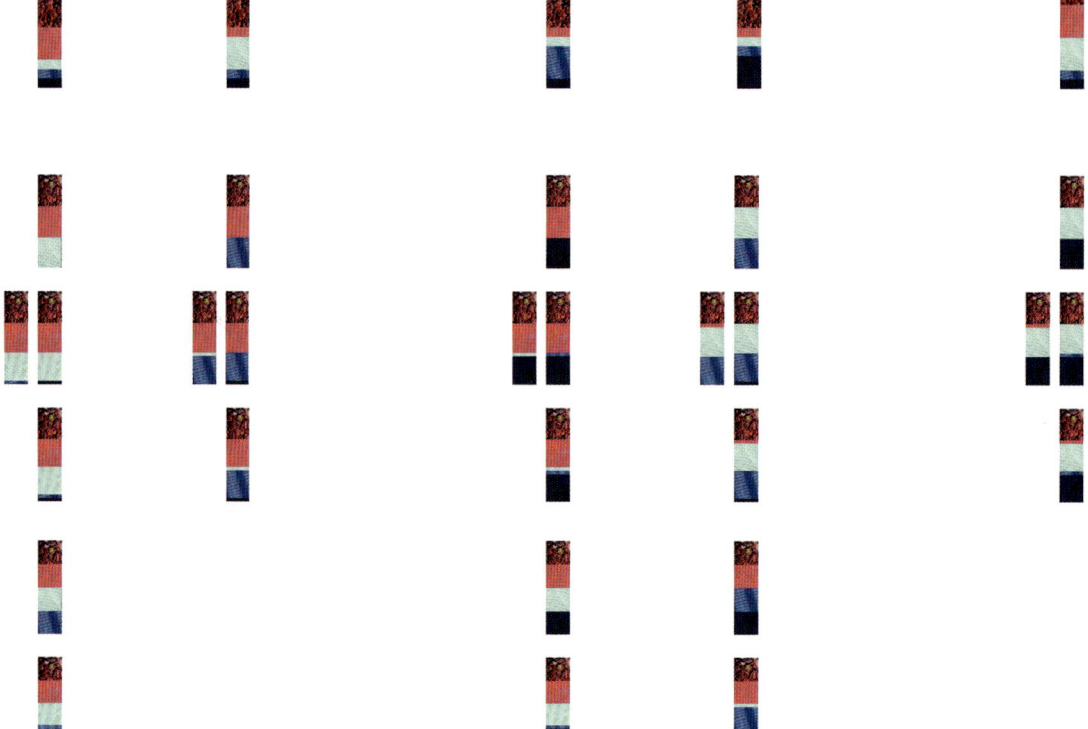

FIGURE 33 Overview, less than 35 percent.

rounding land for vegetation, resulting in patches dominated by buildings. Thus, these patterns appear to reflect social, economic, infrastructural, and design legacies.

Woody Vegetation, Herbaceous Vegetation, and Buildings
Patches that represent this suite are either smaller and located lower in the watershed or larger and located in the middle section of the watershed. The smaller patches reflect alterations to the rowhouse fabric due to abandonment and demolition in such older neighborhoods. Prior to abandonment, these patches would likely not have had

co-dominance by herbaceous vegetation, as buildings would have occupied most of the block. The larger patches in this suite reflect big suburban developments from the middle of the twentieth century; these developments had uniform lot layouts and intentionally included large swaths of grass and shade trees in the yards. The landscape aesthetics of these large suburban developments were shaped by both formal regulations and societal norms. Built at a time of increasing personal auto use, these developments were frequently located near highways; in the Gwynns Falls watershed they are concentrated near beltway intersections.

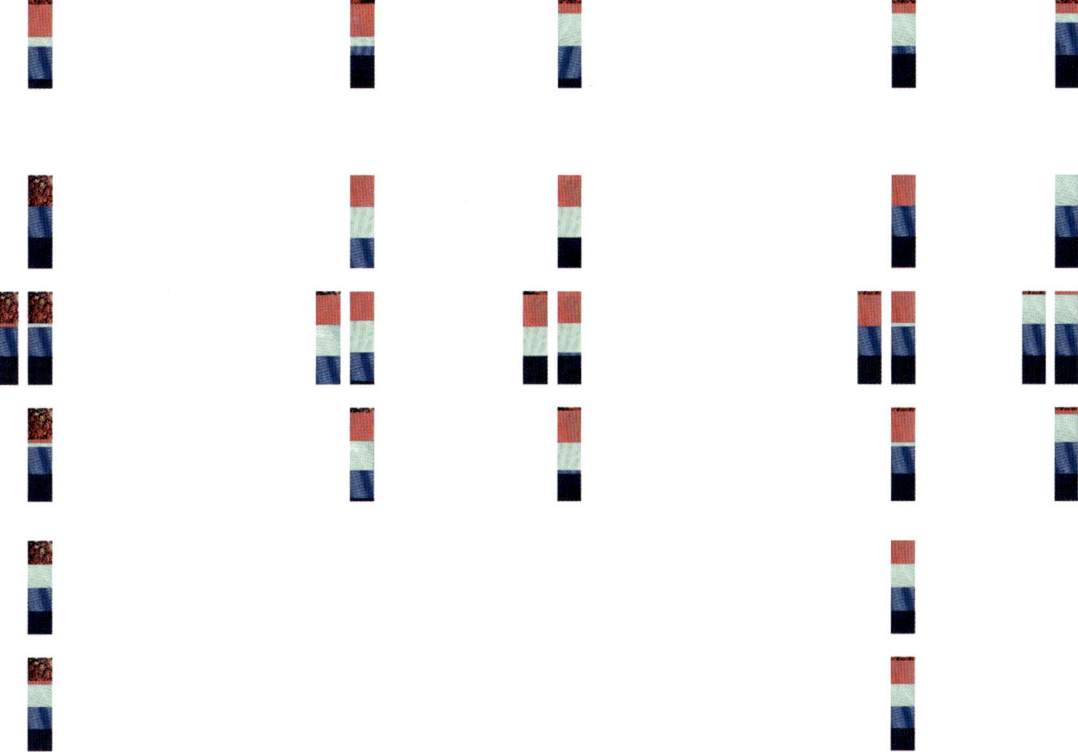

Although these developments are now fifty to seventy years old, and the trees have matured, the original plan and form are still evident. Because of the co-dominance characteristic of this suite, these patches may be susceptible to mutability in the future, in spite of the persistence of land use.

Woody Vegetation, Pavement, and Building Patches in this suite are generally small and distributed across the watershed. Their form is highly variable. For example, some are large commercial and institutional buildings with a lot of trees and parking. Others are old rowhouse neighborhoods that have

experienced abandonment in the inner block when substandard alley houses were removed, or abandoned buildings on the main streets were removed. Such inner block areas were then converted to recreational areas or small parks. Vacant lots on the larger streets often supported trees that self-seeded and frequently reflect absence of management of grassy areas. Idiosyncratic management by single institutional owners, lack of management in marginal parts of commercial properties, or abandonment apparently yield high mutability in patches of this suite.

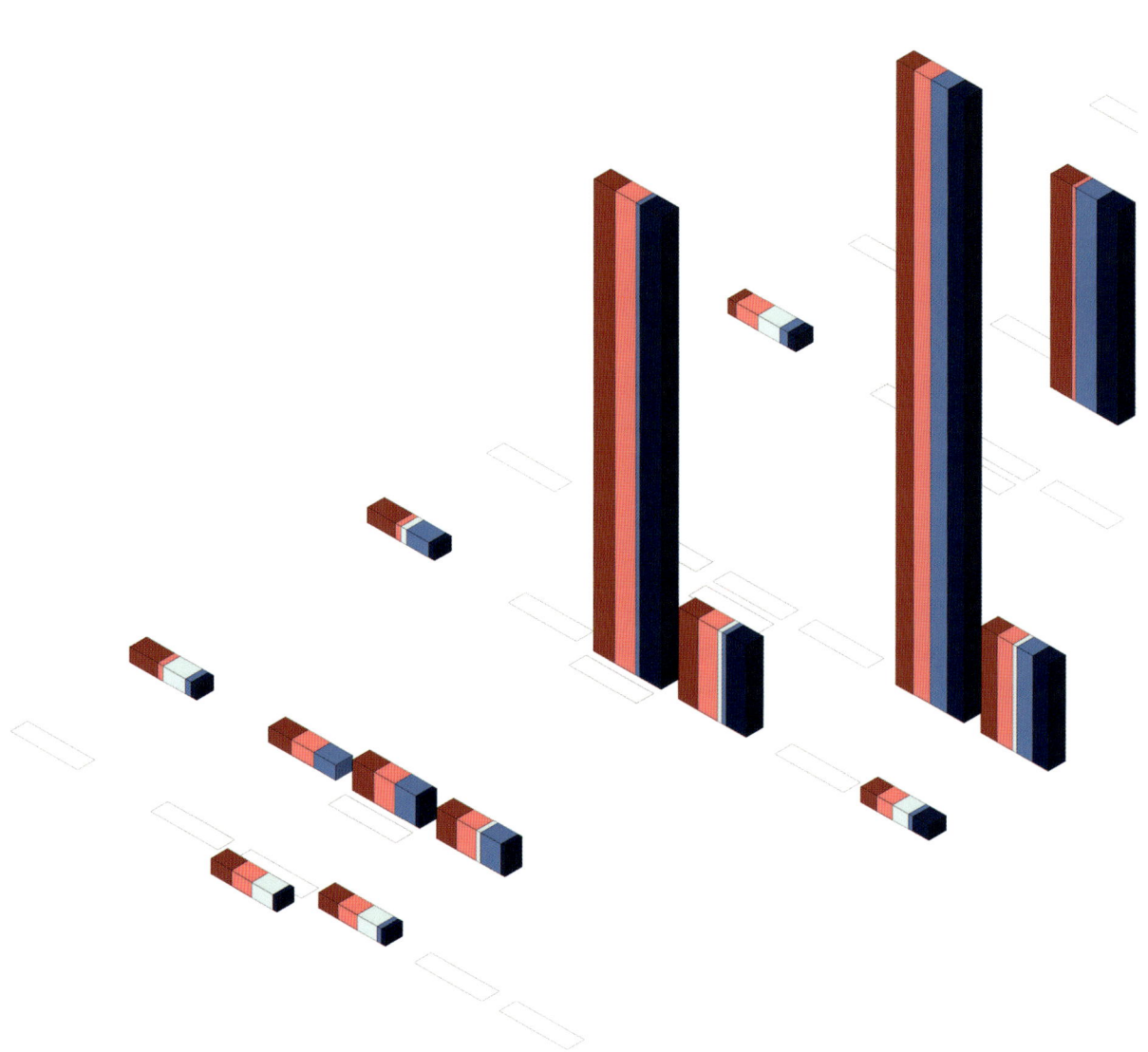

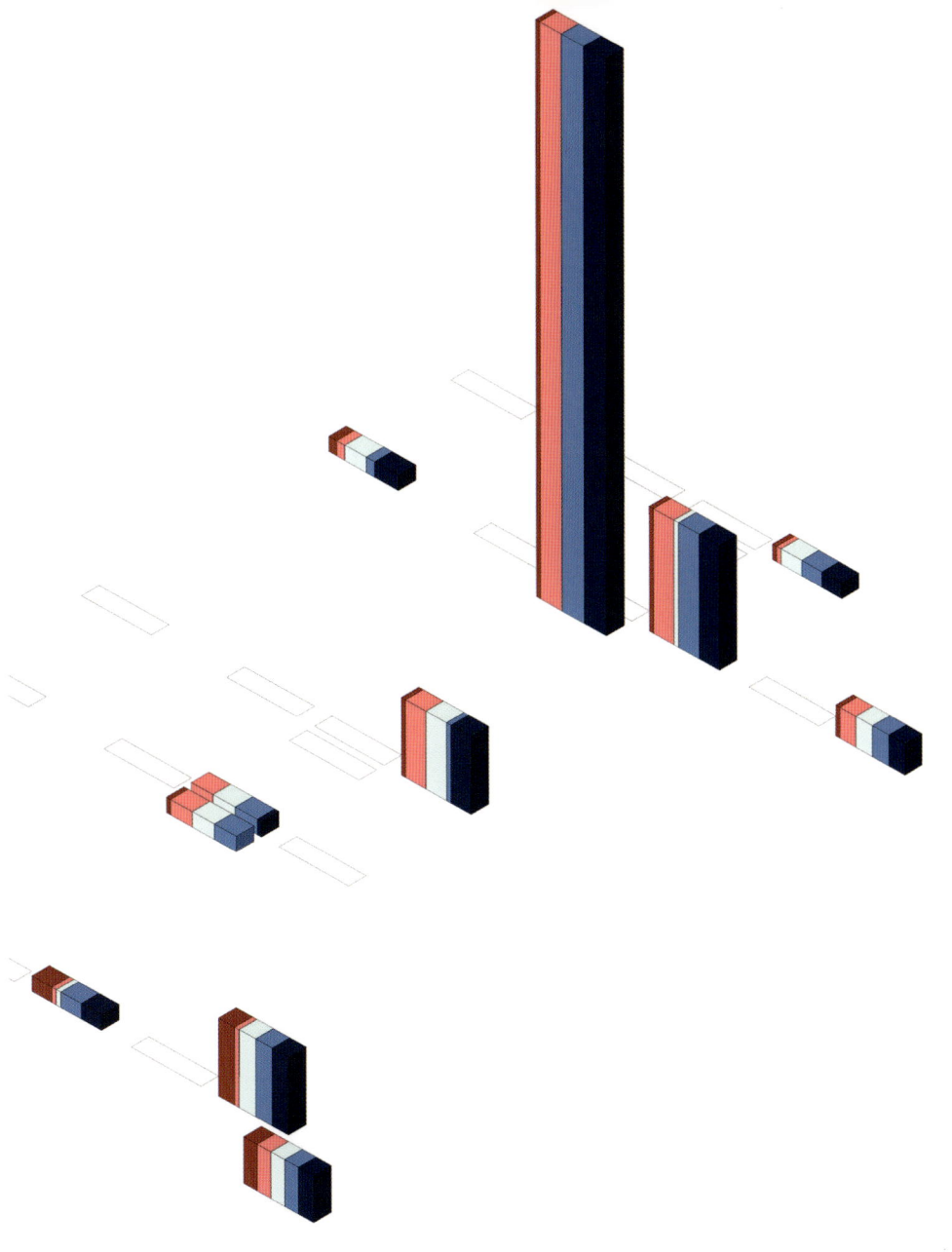

FIGURE 34 Overview, less than 35 percent.

FIGURE 35 Woody vegetation, herbaceous
vegetation, and buildings.

FIGURE 36 Woody vegetation, pavement, and
buildings.

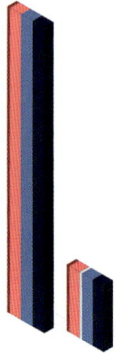

FIGURE 37 Herbaceous vegetation, pavement, and buildings.

FIGURE 38 Woody vegetation, herbaceous
vegetation, pavement, and buildings.

FIGURE 39 Woody vegetation, herbaceous
vegetation, bare soil, pavement, and buildings.

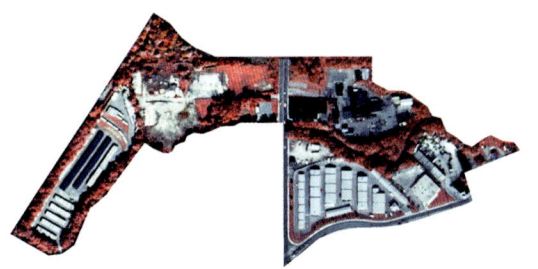

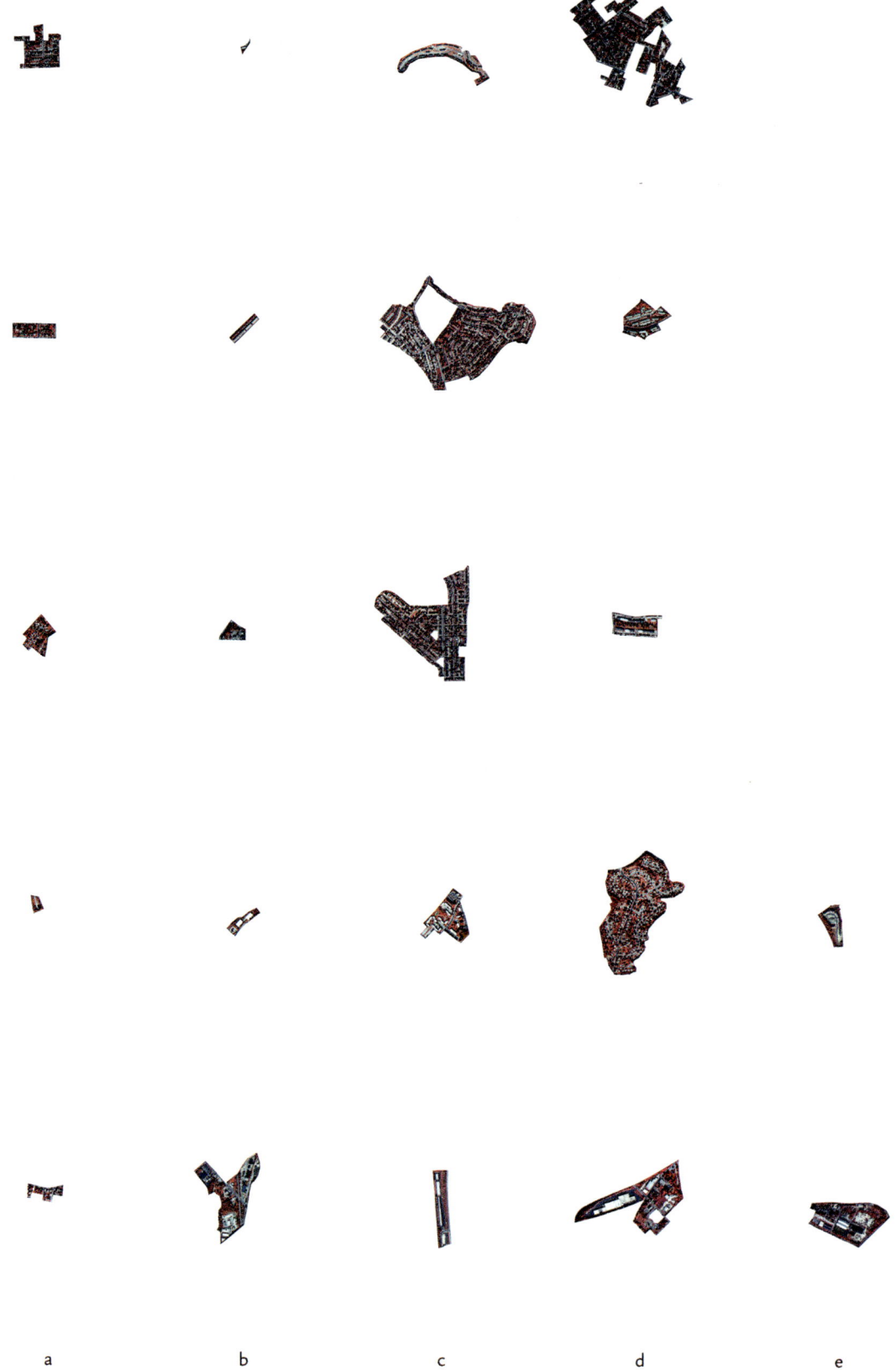

a   b   c   d   e

FIGURE 40 Patch sample contact sheet.
a. Woody vegetation, herbaceous vegetation, and buildings b. Woody vegetation, pavement, and buildings c. Herbaceous vegetation, pavement, and buildings d. Woody vegetation, herbaceous vegetation, pavement, and buildings e. Woody vegetation, herbaceous vegetation, bare soil, pavement, and buildings

Herbaceous Vegetation,
Pavement, and Building
Patches of this type are mostly dense residential areas in the form of condominiums or rowhouses with either shared parking or a dense street and alley grid, making pavement co-dominant. The patches are distributed throughout the watershed, but the larger ones are found in the lower part. Because of high density, buildings and paved surfaces occupy most of the parcels, leaving little space for trees, and allowing small grass borders to cumulatively achieve co-dominance. A few of these patches are commercial or institutional, but regardless of use they share the same form.

Woody Vegetation, Herbaceous Vegetation,
Pavement, and Building
In the watershed map it can be seen that patches that are co-dominated by woody and herbaceous vegetation, pavement, and buildings tend to be concentrated in the mid and lower watershed. They are of varying sizes and well dispersed relative to each other. They range from mature suburbs to institutional, office, and commercial uses. Such mature suburbs support large trees and have parcels small enough to contain a dense street grid. Institutional and commercial areas in this suite also contain large trees and paved areas for parking. Because these are all established landscapes, there is no bare soil in this suite.

Woody Vegetation, Herbaceous Vegetation,
Bare Soil, Pavement, and Building
Patches that are co-dominated by all five land cover elements are extremely rare in the watershed. There are only three examples present, and because of the co-dominance by bare soil, these patches contain areas that are either partially under construction or heavily used, preventing the growth of vegetation. Bare soil may indicate high potential for mutability of these patches in the future.

An atlas is traditionally used for exploration. This one is also for speculation. As a patch rather than a universal atlas, this book's journey is nonlinear and complex, as it explores a specific area of Baltimore in great detail, not as a narrative journey but as a way of understanding urban systems. This atlas is not just a series of maps, but it is a series of partial maps, and charts that are derived from a GIS data base/model. The journey is not just an exploration of a territory, but a speculation on urban futures.

An Atlas for Exploration

An atlas is, in part, a collection of maps that guide one through a territory. People usually think of atlases as being complete and detailed. But the atlas we have produced is both systemic and patchy. It is not intended to be a finalized or singular roadmap where all is certain and laid out in plain sight. Rather, it is an atlas composed of tables, suites of swatches, graphs, and patchy maps to show the territory in a new way. This atlas won't take you from point A to point B in

FIGURE 41 An example of land that has four co-dominant land cover elements: in this case, woody vegetation, buildings, herbaceous vegetation, and pavement.

the most direct or quickest way. In fact, it is intended to slow the reader down, so that one can see urban environments in a fresh and detailed way.

Our atlas employs a bit of purposeful disorientation to provoke the user to understand the Baltimore city region anew. This Patch Atlas sees the city as a patchwork that can be conceptually disassembled and reassembled in various mental ways. The journey therefore proceeded from homogeneity to heterogeneity, directed by the periodic table. It jumps not north to south or east to west but from one patch suite to another. By identifying and focusing on the urbanization patterns and fragments that make up the coverage of an urban area, we hope to have exposed relationships that may not be apparent when looking at the whole or passing by on the path from place to place.

Most important, this diffracted way of seeing raises questions about the relationships that do and must exist across a complex spatial logic of a patch mosaic. The questions call out different scales of spatial observation; they beg to know the changes in particular urban patches over time, including their origin, transformation, migration, and even disappearance; they inquire

Ten Thoughts

1. We have learned over seventeen years in the Baltimore Ecosystem Study to respect the differences between ecological and design thinking. We are all patient listeners, waiting for concepts to form in our own heads and as articulated by our colleagues. What we found is that within both research and practice, similar actions occur; however, the sequence of model, projection, and understanding differs. Therefore, we have worked to correlate these different methodological sequences and thought processes in nonlinear ways.

about the shape and shift in the entire mosaic of patches. And finally, holding up the patches individually and in related groups for considered examination makes us wonder about the rich array of human, institutional, and ecological agency that underlies their characteristics and responds to their evolution. The extraordinary diversity of HERCULES patches belies the simplicity of the land use maps that most cities use for planning decisions. Instead, the Patch Atlas is an exploratory tool, where the potential change imminent within each patch outweighs the possibilities of planning the whole.

An Atlas for Speculation and Inclusion

The second journey we have set out for you is one that leads to our new approach for a speculative and inclusive understandings of the complexities of urban land cover. The speculative journey was motivated by a desire to improve the way that spatial heterogeneity and hybridity of urban regions is described and classified in order to further the agency of social choice and the power of inclusive design. When we started our own exploration of urban heterogeneity, we discovered that we needed a new tool for capturing and designing urban heterogeneity—

2. We have found that even when ecologists and designers use the same words, they may mean very different things. So, we became open to expanding our closely held definitions and to be comfortable with another disciplinary understanding of a word or term. Key words stimulated shared theoretical frameworks that bridged ecology and design. Such seminal words include spatial heterogeneity, patch dynamics, watershed framework, and metacity theory.

the specific patchiness that is a driver, an outcome, and a potential of the urban/suburban/exurban mix in Baltimore.

To satisfy this need, we developed HERCULES—the High Ecological Resolution Classification for Urban Landscapes and Environmental Systems. This approach to urban land cover focuses on just that—the cover—and not on more familiar urban classification systems that emphasize land use. Land use categories do imply certain physical arrangements of things that cover the ground, but they primarily are motivated to describe how people use those covers. The categories in land use/land cover classifications include such things as agriculture, forest, residential, commercial, industrial, transportation, barren, and water. For many purposes these are fine and good. But for seeing cities in a new way, they fall short. Taking an aerial image as pixels to be classified as either agriculture, or forest, or urban, and so on, runs the risk of reinforcing the traditional urban versus natural, cultural versus wild divides that are so familiar and unfortunately so comfortable.

Urban ecological science, social-ecological system science, human geography, and design all conceive of urban/suburban/exurban/rural systems as

3. Both ecology and design are creative and inclusive acts engaged in the task of fostering environmental understanding and action. We do not adopt a goal of scientific management or prediction of the performance of green spaces in cities. Instead, we engage in design with people for cities. We think, design, model, and act within the biophysical world and speculate on collective cultural understandings of urban ecosystems through sensory aesthetic experience, as well as thought experiments.

hybrid social-ecological territories. These approaches all set aside or seek to overcome the familiar and intransient human-versus-nature divide. They have introduced a way of seeing city systems, and human ecosystems in general, as hybrids. And hybrids are not divisible entities. Their sources and origins are constitutionally inseparable. One can extract neither a horse nor a donkey from a particular mule. HERCULES asks, what is the nature of urban spatial heterogeneity as a constitutional hybrid of social processes and products and of biophysical processes and products. HERCULES seeks to describe urban heterogeneity as socially and biophysically co-produced.

The journey of creating this atlas aims not only to represent urban form as hybrid, but it is itself a hybrid journey that aims to inquire into and alter the processes of knowing and (re)making urban areas. Just as urban patchiness is co-produced by social and biophysical processes, so too was this atlas co-produced by an intense and long-term interaction, which melds the insights of several disciplines and professions. Two of us are scientists and initial authors of the HERCULES approach. And two of us are urban designers—a landscape architect and an architect, who articulated the value of exploring HERCULES for their theory, design studio teaching, and urban design practice.

4. Allying ecology and design together is a public good. As the Ecological Society of America has stated (Sayre et al. 2013), there is no time for the normal, linear scientific process of waiting for empirical proof of a theory to act in a world waiting for solutions. Just as sustainability is an endeavor of simultaneous action by multiple parties, so do theory, discussion, models, designs, and experiments need to occur in parallel, always led by diverse community actors.

Although scientists exhibited creativity in their own realm in producing the new hybrid classification system of HERCULES, they produced an academic depiction of the system. Enter the design specialists and theorists, and a dialog among professions produced—co-produced actually—a novel way to explain and visualize the classification. The urban designers saw the classification as representing a grouping of possible types of patches, and demonstrated the value of visualizing patches as individual shapes and as pattern maps, systematically organized at the same scale and aligned in order to show similarities and differences among suites. The designers added a new dimension of speculation. The feedback between science and design was multifaceted, involving data tables, exploratory drawing, discussion, mutual confusion, drawing again, and still more discussion. We distilled ten thoughts that reflect on this experience. That seventeen years of dialog and drawing is the journey that has led to the atlas you have been exploring.

The format of the patch drawings in the atlas is significant. We have chosen a more creative rather than strictly cartographic strategy for representation. This means that outside of the individual patches details of neighborhood and of boundary are not shown. Neither do we show such conven-

5. Both ecology and design are creative and inclusive acts, and they also employ scientific rigor. Design is not an intuitive application of style on a pre-engineered and scientifically proven system. Designs must be validated, but the measuring instruments are blunt. Emerging practices of community-informed research and inclusive design provide sharper, finer grain, and faster feedback loops to inform design and ecology experiments in the making.

tions as latitude and longitude, a scale bar, north arrow, watershed boundary, or city/ county border. We do not assert that these things are not important, but rather we have made this choice to focus on the heterogeneity that each patch type and suite of patches represent and the latent potential to reimagine the whole through the particular nature of the parts. If anything, these aesthetic choices point to research questions and other design options that must be explored elsewhere. The focus here is on the heterogeneity within patches as the fundamental concern among an always-present context of flows across adjacent patches as

well as patterns across the watershed. Other scales of heterogeneity appear as background or open questions in the narrative about each patch suite.

The act of taking the gradient of land cover elements and breaking them into suites continues to be a focus of discussion. The suites, described here as dominant, co-dominant, and plural, are the result of a design strategy employed to make the novel and gigantic patch data base intelligible at a glance. While there is no inherent significance of a co-dominant patch changing into a plural patch or alternatively into a dominant patch, the suites became a useful way

6. Societal and political forces devalue both fields when they are put in separate boxes of the creative and the rational, inducing competition rather than cooperation between science and humanities. The funding models of the National Science Foundation and the National Endowment for the Humanities, as well as academic promotion and advancement strategies, rigorously police these disciplinary boundaries, hindering invention. We aim not to serve these instruments of power but rather to serve the communities we work with.

to communicate among us, and may perhaps become a stimulus for future speculative and inclusive design inquiry. Was something lost in this decision? For example, if you divide land covers into a number of categories and make a cutoff between major categories of covers you will necessarily get an arrangement—a geometry. There is a geometrical structure to the atlas, and a mathematician who studies combinatorics might come up with a different structure. Or an empirical scientist might be interested to create bar graphs and align what is geometrically possible versus what is actual, in order to expose interesting constraints.

A next step might be to employ the atlas as a speculative and inclusive design tool. For example, a participatory process might ask: What new mixes of forest and built fabric do residents prefer, where, and why? What knowledge is required to intentionally reshape heterogeneity in a patch mosaic? What ecological questions arise? And how might these inform ecosystem science research and experiments? Another possible next step might be to employ the atlas as a design modeling exercise. How do different geometries motivate urban design practice at different scales? How does the degree of contrast in urban form between adjacent

> *7. It is important to support new forms of research, drawing, designing, making, monitoring, and communicating. We engage a new type of creative labor that is beyond rendered surfaces usually generated by design, where process and individuals are lost, masked, and invisible. We aim to make the work of both science and design inclusive, performative, and visible.*

patches influence or motivate urban design ideas?

The Atlas as a Work in Progress

The journey is not over. But the atlas does lay out some milestones along the way, and points to future exploration. Here are some of our key insights, and stimuli for moving on with the atlas in hand.

The maps in the atlas are just one marker in time. But spatial changes in Baltimore are not over. And the spatial patterns we have displayed here have a long history emerging from the land, from trade, from industry, from segregation and migration, from

globalization, from the deployment of political power, and the exercise of social innovation, among many others. The atlas is an invitation to explore the future potential of the region through its patches, as well as a deeper understanding of its past.

The Patch Atlas is not only a compelling representation. Of course, we hope it is visually attractive and creatively inspiring, but importantly, the images are rooted in data. In this case, the data come from a set of false-color infrared photographs taken in 1999. It is a fine-scaled representation of the cover of the Gwynns Falls watershed on a particular date in October of

8. *Both fields suffer when just responding to a marketplace based on fear or security as a motivator for action. We design within the contested politics of the public sphere. We creatively link empirical approaches, social and physical interactions, environmental and social justice, and aesthetic and sensorial pleasure in a context of diverse community participation in response to rapid climate change and structural inequity.*

that year. Heads-up digitization was used to delimit patches, and the covers of all types within each patch were estimated in defined numerical suites. Thus, there are geo-referenced quantitative data that underlie the patch maps throughout the atlas. It is possible to query the data base to focus on particular groups of patches or on particular places in the watershed. Therefore, this product is scientifically rigorous and, because of its visual abstraction, it may also invite broader engagement by various audiences. The Patch Atlas is designed to provide the reader a sense of interacting with the HERCULES data base.

The HERCULES approach uses generalized cover elements—buildings, woody and herbaceous vegetation, paved and bare surfaces. These elements can be discerned in any landscape, though it was originally motivated by a need to characterize urban and urbanizing areas. Hence the approach is generalizable and adaptable to various situations and places.

By applying HERCULES in Baltimore and representing this as an atlas, two logics have been exposed. The first is the mathematics of combining the five basic cover types into swatches based on quantitative cutoffs of percent cover. This logic using the

9. Trust is essential, both between disciplines and with the communities in which we are engaged. Urban ecological design research and practice is not an excursion into new knowledge only to come back to a disciplinary home again refreshed. It is not for inspiration or stimulation alone. It is a commitment to change the foundations of understanding. It includes failure—from which lessons emerge.

cutoffs and rules described earlier in the atlas shows the potential kinds of heterogeneity of cover that can exist in our watershed. Following this combinatorial logic we have created a "periodic table," with obvious reference to the regularities of the chemical elements. But there is a second logic as well, which has documented an array of actual patches that exist in the Gwynns Falls watershed. The patterns these actual arrays display may be a "signature" of this 17,150-hectare part of the Baltimore region, or of different parts of the watershed. Comparing signatures of actual patches from different places and times may turn out to be

a powerful tool in urban ecological design theory and practice.

Focus on pattern alone is a sterile exercise. In fact, our intent in even creating the HERCULES approach was to provide a clean and unambiguous data set against which to relate functions or processes. In science, a core pursuit is to discover and evaluate causal relationships that may exist between structure or pattern and function or process. We realize that some of these terms will be problematic to some readers, and so we offer the caution here that we do not see intentionality in all pattern-process relationships, although urban systems and urban

10. A metacity approach emphasizes exploring a mosaic of patchy, local action. Diverse communities of ecological design practice are increasingly coming together to make a difference. Large heterogeneous spatial systems are aggregated through patch and watershed frameworks within this wider perspective and expanded time frame. Theory creates a meta-methodology for transdisciplinary understanding and action converging in inclusive local community projects within larger scale frameworks.

design do embody clear intentionality. They also involve complex emergence, accidental effects, and indirect lags and legacies. Heterogeneity can be a key driver of all such outcomes and effects.

We have aspirations for the Patch Atlas. Among us, we have many reasons for involvement in urban areas. These range from the joy of understanding cities as complex, historically situated systems, to the delight of experiencing the heterogeneity in Baltimore as we walk or drive about, to the need for framing rigorous scientific observations and models, to the use of improved models of urban form as a basis for form/

function analysis, to identifying places in which design initiatives may be most needed or impactful, among many others. We intend to use the atlas as a tool for our own continued interdisciplinary dialog and practice. We hope that other groups, whether seasoned professionals or beginning students, can use it for their own dialogs and designs. Even better, if this atlas stimulates other groups to pursue an open, inclusive approach to understanding heterogeneity and improving urban research and practice beyond "land use" in their own cities, that would be our greatest success.

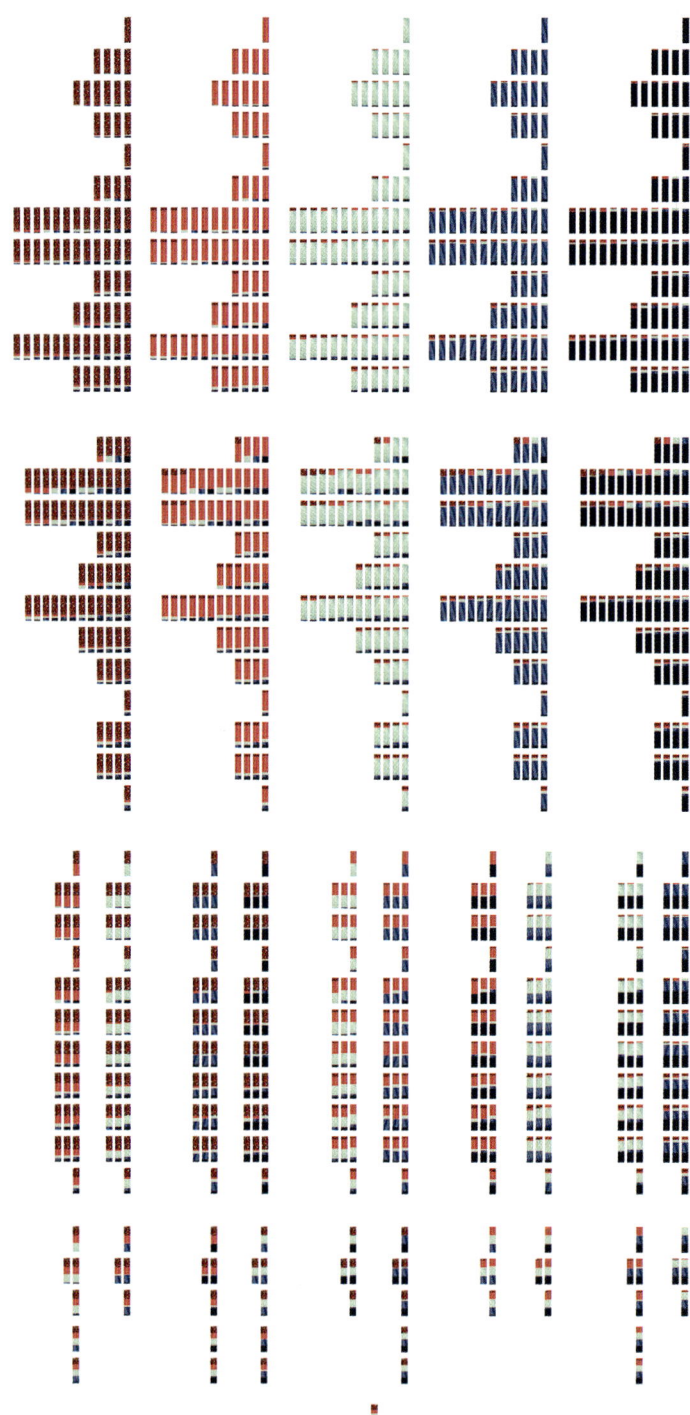

FIGURE 42 The periodic table, with all possible swatches, left, and the full graph of actual patch types, right. They are shown side by side and at the same scale so the graph "signature" of the Gwynns Falls watershed can be seen as a partial and specific arrangement of swatches or patch types that are actually present. Going forward we aim to create and compare "signatures" from other urban areas.

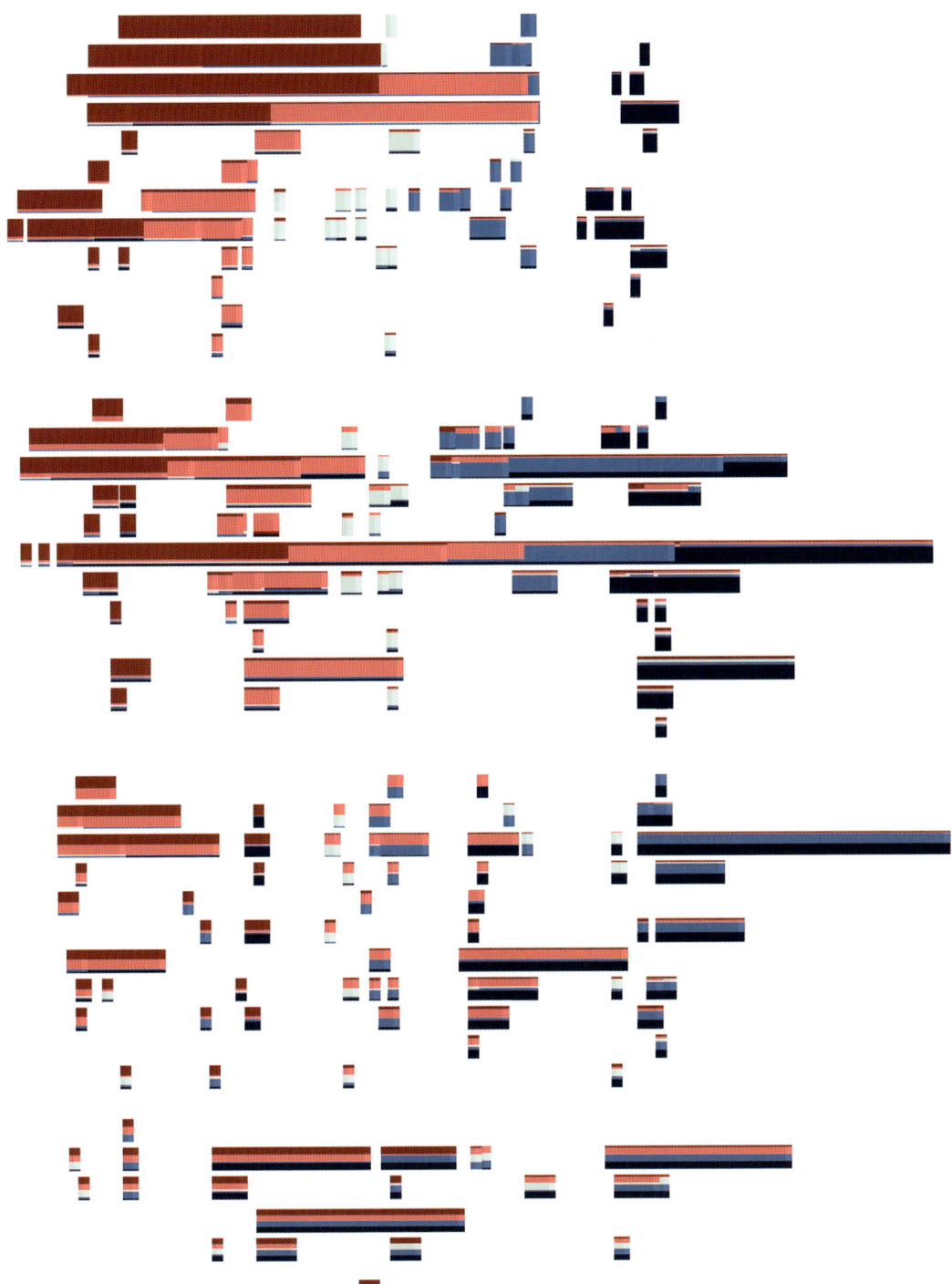

FOR FURTHER READING

Anderson, J. R., E. E. Hardy, J. T. Roach, and **R. E. Witmer.** 1976. Land Use and Land Cover Classification System for Use with Remote Sensor Data. Geological Survey Professional Paper 964. US Government Printing Office, Washington, DC. This report describes the original "industry standard" land classification system developed for the United States. Using the then emerging technology of remote sensing, the scheme proposes a hierarchical classification that begins by sorting "urban or built-up land" from forests, or agricultural lands, for example. This scheme has been widely adopted as a basis for international classification methods as well.

Cadenasso, M. L., S. T. A. Pickett, and **K. Schwarz.** 2007. Spatial heterogeneity in urban ecosystems: Reconceptualizing land cover and a framework for classification. Frontiers in Ecology and the Environment 5:80–88. Here the HERCULES system is described in detail, compared to other classification approaches, and linked to focus on understanding the relationships between pattern and process.

Clay, G. 1973. Close-Up: How to Read the American City. University of Chicago Press, Chicago. An eloquent account of how a prominent urbanist, and one-time editor of Landscape Architecture magazine, observed at close hand and fine scale the complexity of American cities. Grady Clay's books, newspaper columns, and radio spots serve as a humanistic context of the technical approach of HERCULES.

Dramstad, W. E., J. D. Olson, and **R. T. T. Forman.** 1996. Landscape Ecology Principles in Landscape Architecture and Land-Use Planning. Island Press, Washington, DC. Intended as a handbook for designers and planners, this brief book takes the classical principles of landscape ecology, such as the patch/corridor/matrix approach, and distills them in terms useful for application.

Ellis, E. C., R. G. Li, L. Z. Yang, and **X. Cheng.** 2000. Long-term change in village-scale ecosystems in China using landscape and statistical methods. Ecological Applications 10: 1057–73. A fine-scale and highly resolved system, initially developed for Chinese village life, but applicable much more broadly. This approach combines human use and cover into a spatially and functionally refined system of identifying patches or "ecotopes" that are associated with human settlements in rural and urban situations.

Geddes, P. 1915. Cities in Evolution: An Introduction to the Town Planning Movement and the Study of Civics. Williams & Northgate, London. A classic work from the dawn of regional planning. Geddes was at pains to link evolutionary processes, regional gradients of natural potential and land use, and the social with the biological. As such, he stands as a pioneer of the hybrid approach to land that HERCULES attempts to operationalize at fine to coarse scales and apply within cities as well as urban regions. Meller (1990), cited below, places Geddes in a contemporary context.

Gottdiener, M., and **R. Hutchison.** 2011. The New Urban Sociology. 4th edition. Westview Press, Philadelphia. A key summary of the role of spatial heterogeneity in social patterns and processes. A reminder of the way that social data and concerns can be linked with the spatial approach taken in this Patch Atlas.

Grove, M., M. L. Cadenasso, S. T. A. Pickett, G. Machlis, and **W. R. Burch Jr.** 2015. The Baltimore School of Urban Ecology: Space, Scale, and Time for the Study of Cities. Yale University Press, New Haven. A conceptual and historical overview of the social-ecological research program in the Baltimore region from which HERCULES emerged. A useful context for the origin and the application of the patch approach to urban systems.

Held, D. 1980. Introduction to Critical Theory. University of California Press, Berkeley. An introduction to, and evaluation of, critical theory.

Marshall, V. 2013. Patch Reflection. The Nature of Cities. https://www.thenatureofcities.com/2013/04/14/aerial-reflection-for-urban-ecology/. An approachable description of how to detect and map patches, based on the author's extensive work in Asian cities.

McGrath, B. P., editor. 2013. Urban Design Ecologies. John Wiley and Sons, Ltd, Hoboken. An anthology of original and collected essays that explore the manifold relationships between design and ecological understanding. A conceptually open framework emerges from this collection.

McGrath, B. P., V. Marshall, M. L. Cadenasso, J. M. Grove, S. T. A. Pickett, and **J. Towers,** editors. 2007. Designing Patch Dynamics. Columbia University School of Architecture, Planning and Preservation, New York. This book is the result of three years of interaction among the authors in conducting ecological design studios at Columbia University. Chapters describe the basic ecological concepts of patch dynamics, boundaries and edges, ecosystems and watersheds. These chapters are complemented by those that focus on design as an expanded field and is therefore well suited to break down disciplinary barriers. Student projects that illustrate the issues and successes of crossing the design/ecology suture are presented.

McGrath, B., and **S. T. A. Pickett.** 2011. The metacity: A conceptual framework for integrating ecology and urban design. Challenges 2:55–72. Although the metacity is mentioned in the Patch Atlas, it is not one of the main included topics. This paper provides a more extensive and illustrated introduction to the concept of the metacity.

McHarg, I. L. 1969. Design with Nature. Natural History Press, Garden City, NY. A pioneering classic at the contact between ecological, social, and design understandings. HERCULES is in a sense, a more hybrid fruit of McHarg's approach of optimizing discrete maps of the different aspects of landscapes undergoing development.

Meller, H. 1990. Patrick Geddes: Social Evolutionist and City Planner. Routledge, New York. This book places the life and work of Geddes in its contemporary context.

Pickett, S. T. A., M. L. Cadenasso, and **B. McGrath,** editors. 2013. Resilience in Ecology and Urban Design: Linking Theory and Practice for Sustainable Cities. Springer, New York. This book examines the evolving connections between ecology and design, by collecting the theories, urbanistic philosophies, design approaches, ecological and social patterns together in one volume to facilitate further growth at that boundary.

Pickett, S. T. A., M. L. Cadenasso, J. M. Grove, E. G. Irwin, E. J. Rosi, and **C. M. Swan,** editors. 2019. Science for the Sustainable City: Empirical Insights from the Baltimore School of Urban Ecology. Yale University Press, New Haven. Key findings and insights from over two decades of research, education, and community engagement in the acclaimed Baltimore Ecosystem Study.

Pickett, S. T. A., and **P. S. White,** editors. 1985. The Ecology of Natural Disturbance and Patch Dynamics. Academic Press, Orlando. A classic exposition of the role of natural disturbance in creating spatial patchiness in ecological systems. The approach is applicable to key processes within urban systems, whether of natural or social origin.

Pickett, S. T. A., J. Kolasa, and **C. G. Jones.** 2007. Ecological Understanding: The Nature of Theory and the Theory of Nature. 2nd edition. Academic Press, San Diego. A philosophy of ecology, describing the structure, content, and development of scientific theory in terms that are particularly relevant to ecology. The book was the first to correct the neglect of ecology by most philosophers of science. The theoretical perspective is implied by the structure of HERCULES presented in the Patch Atlas.

Shane, D. G. 2005. Recombinant Urbanism: Conceptual Modeling in Architecture, Urban Design and City Theory. John Wiley & Sons, London. This important urban design history and theory text argues that cities develop and grow according to three models: enclaves, armatures and heterotopias. It provides important insights into how cities adapt and grow over time.

Spirn, A. W. 1984. The Granite Garden: Urban Nature and Human Design. Basic Books, New York. This classic work takes the environmentally sophisticated view begun by McHarg, and applies it to existing cities and urban areas. In doing so, it points the

way toward a more thoroughgoing hybrid-
ization of biophysical and social phenomena
that can be useful for design. In addition,
issues of social and environmental justice
are clearly identified and called out for miti-
gation in this book.

Steiner, F. R., G. F. Thompson, and **A. Car-
bonell,** editors. 2016. Nature and Cities:
The Ecological Imperative in Urban Design
and Planning. The Lincoln Institute of
Land Policy, Cambridge. One of the mature
flowerings of ecologically oriented design,
showing how design can employ ecological
knowledge effectively.

**Sayre, N. F., R. Kelty, M. Simmons, S. Clay-
ton, K.-A. Kassam, S. T. A. Pickett,** and **F. S.
Chapin.** 2013. Invitation to Earth stewardship. Frontiers in Ecology and the Environment 11:339. Earth Stewardship is an initiative of the Ecological Society of America. It
seeks to broaden and deepen the connections between ecology and various other
disciplines and professions that must be
engaged together in improving the sustainability of the Earth. Design is one of the professions and perspectives that is a key part
of stewardship partnerships.

Zhou, W., M. L. Cadenasso, K. Schwarz,
and **S. T. A. Pickett.** 2014. Quantifying spatial
heterogeneity in urban landscapes: Integrating visual interpretation and object-based
classification. Remote Sensing 6 (4): 3369–
86. A methodological demonstration of how
to apply the HERCULES system using new
analytical techniques. The work shows how
the hybrid approach to patch description
has recently advanced as a method. This
complements the theoretical and conceptual
approach emphasized in the Patch Atlas.